Francis Orpen Morris

A History of British Butterflies

Vol. 3

Francis Orpen Morris

A History of British Butterflies
Vol. 3

ISBN/EAN: 9783743377929

Manufactured in Europe, USA, Canada, Australia, Japa

Cover: Foto ©berggeist007 / pixelio.de

Manufactured and distributed by brebook publishing software (www.brebook.com)

Francis Orpen Morris

A History of British Butterflies

A

HISTORY

OF

BRITISH BUTTERFLIES.

BY

THE REV. F. O. MORRIS, B.A.,

MEMBER OF THE ASHMOLEAN SOCIETY.
LIFE MEMBER OF THE BRITISH ASSOCIATION FOR THE ADVANCEMENT OF SCIENCE, ETC., ETC.
Author of a "History of British Birds," Dedicated by Permission to Her Majesty the Queen,
a "Natural History of British Moths," etc., etc.

SIXTH EDITION,

WITH SEVENTY-TWO PLATES,

COLOURED BY HAND.

LONDON:
JOHN C. NIMMO, 14, KING WILLIAM STREET, STRAND.
M DCCC XC.

TO

THE HON. MRS. MUSGRAVE,

BISHOPTHORPE PALACE,

THIS VOLUME

IS

BY HER PERMISSION

MOST RESPECTFULLY DEDICATED

BY

HER OBLIGED AND OBEDIENT SERVANT

THE AUTHOR.

CONTENTS.

	PAGE		PAGE
Swallow-tail	1	Purple Hairstreak	81
Scarce Swallow-tail	4	Green Hairstreak	83
Brimstone	7	White-W Hairstreak	85
Clouded Yellow	9	Black Hairstreak	87
Pale Clouded Yellow	13	Brown Hairstreak	89
Black Veined	15	Duke of Burgundy Fritillary	91
Large White	17	Greasy Fritillary	93
Small White	19	Glanville Fritillary	95
Green Veined	21	Pearl-bordered Fritillary	97
Chequered White	23	Small Pearl-bordered Fritillary	99
Wood White	25	Pearl-bordered Likeness Fritillary	101
Orange Tip	27	Weaver's Fritillary	104
Marbled White	29	High-brown Fritillary	106
Wood Argus	33	Dark-green Fritillary	109
Wood Ringlet	35	Queen of Spain Fritillary	111
Gate-Keeper	37	Venus Fritillary	113
Rock-Eyed Underwing	39	Silver-washed Fritillary	114
Small Meadow Brown	41	Large Copper	116
Large Meadow Brown	43	Small Copper	119
Heath Butterfly	45	Brighton Argus	121
Least Meadow Brown	48	Mazarine Blue	123
Arran Argus	50	Large Blue	125
Scotch Argus	51	Holly Blue	127
Small Ringlet	53	Little Blue	129
Silver-bordered Ringlet	55	Silver-studded Blue	131
White Admiral	56	Common Blue	134
Red Admiral	58	Clifden Blue	137
Peacock	61	Chalk Hill Blue	139
Large Tortoise-shell	63	Brown Argus Blue	142
Small Tortoise-shell	65	Grizzled Skipper	145
Camberwell Beauty	67	Dingy Skipper	147
Comma	70	Large Skipper	149
Albin's Hampstead Eye	73	Silver-spotted Skipper	151
Painted Lady	74	Small Skipper	153
Scarce Painted Lady	76	Lulworth Skipper	155
Purple Emperor	78	Spotted Skipper	157

PREFACE

TO THE FIRST EDITION.

An instinctive general love of nature, that is, in other words, of the works of God, has been implanted by Him, the Great Architect of the universe—the Great Parent of all—in the mind of every man. There is no one, whether old or young, or of whatever circumstances or rank in life, who can look without any feeling or emotion on the handiworks of Creation which surround him—who can behold a rich sunset, a storm, the sea, a tree, a mountain, a river, a rainbow, a flower, without some degree of admiration, and some measure of thought. He may, indeed, for the time, or for a moment, be engrossed by some worldly care, or some other subject, some remembrance of the past, or anticipation of the future, but this cannot always be the case, and whenever the mind is relieved from that overpowering feeling, the spontaneous thoughts which originate in the love of nature, will be sure to arise in his soul.

Whether indeed in some there is more than a general feeling of this kind; whether all, if opportunities had been afforded to them, and had been afforded to them in good time, would have found that especial delight which others find in the more intimate study of this or that branch of Natural History, whether it may have been only the pressure of different and altogether necessary thoughts that has pre-occupied the mind, and taken away, or, rather, set aside, these which would otherwise have naturally found favour with it, I will not take upon me to determine; but thus much I can and do say, because I can say it of and for myself, that with me, in this sense, the universal includes the particular—includes every particular that is included under it; for there is no group of the wide-spread family of nature that I do not love to study, and to become more and more intimately acquainted with the members of. They are all the creations of the same wonderful Being—"the hand that made them is Divine!"

And if there be one branch of Natural History which is to me more captivatingly interesting than another, it is Entomology; one which is moreover so easy of full gratification, so compatible with every pursuit, so productive of friendly feeling with others, so amalgamative of the high and low together in perfect amity, so singularly pleasing and delightful in itself. I trust, indeed that I have not forgotten, do not forget, and never shall forget, that I have high and holy duties to perform, to which all else must be subordinate and give way. As a servant of the Church, a minister of the Gospel of Christ, I willingly sacrifice natural wishes to the cause of duty. It is but a few brief moments that I snatch for that which is naturally most pleasing to me. Knowing, however, that these studies are innocent in themselves; that they may, with many, prevent other pursuits which, if followed, would assuredly cause risk of most serious danger; that they add to the amount of human happiness, and that, if used as they always should be, they infallibly lead from the works of Nature up to the God of Nature, in feelings of the holiest adoration and most humble worship, I encourage others to follow them, so far as it may be right for them to do so, and have undertaken, at the request of another, to write the following Natural History of British Butterflies, and to supply particulars which I have felt the want of myself.

There are already other works of a similar kind, which have been extensively and deservedly patronized, and of them it is no part, either of my business or my inclination, to speak: neither is it for me to speak of my own: they have spoken for their authors; let mine, too, speak for me.

F. O. MORRIS.

HISTORY

OF

BRITISH BUTTERFLIES.

SWALLOW-TAIL.

PLATE 1.

Papilio Machaon,	LINNÆUS. DONOVAN. HARRIS.
" "	CURTIS. WESTWOOD. DUNCAN.
Papilio Regina,	DE GEER.
Jasonides Machaon,	HUBNER.
Amaryssus Machaon,	DALMAN.

IN all our judgments of objects of Natural History, comparison and relative proportion must guide us to the result. Compared then with multitudes of the exotic species whose dazzling refulgence, splendid hues, elegant forms, and wonderfully varied and eccentric markings, adorn the hills and valleys of far-distant and tropical lands, which these by themselves alone furnish an abundantly exciting wish to visit, the present, our largest British Butterfly—our finest capture—holds but an humble place—"a Satyr to Hyperion" almost,—a foil by their side to their beauty: but we must not, and we do not, despise our own Swallow-tail.

This fine species is said to be found in various parts of Europe, Asia, and Africa, namely in the whole of the former continent, and in the second, even in Siberia, as also in Syria, Nepaul, Cachemere, and the Himalayan Mountains; Egypt, and the coast of Barbary.

In our own country it has been met with in Yorkshire, near Beverley and Cottingham; in Dorsetshire, by J. C. Dale, Esq., in the parish of Glanville's Wootton: he took twelve specimens there in three days, about forty years ago, but has not seen one since. Also at Cranborne, Wimborne Minster, and on Fifehead Common, in the same county, in former years; so too at Pulborough. In Hampshire, Middlesex, Sussex, Essex, and Kent, in Norfolk at Acle, near Yarmouth, in plenty, and also in meadows at Oby and Thurne, in some years in great abundance, as likewise, Mr. Postans informs me, at Horning Ferry, near the ruins of St. Benedict's Abbey, and at Hoveton; also in Somersetshire, at Weston-Super-Mare. But most of all in Cambridgeshire and Huntingdonshire, where, in the fenny districts, it has been, and even is still, very abundant, though, as those parts are fast being drained, it is to be feared that we may in time lose this most conspicuous ornament of our cabinets. They still exist at Wicken Fen.

The perfect insect is taken from the beginning of May to the end of August. One was bred as early as March, by Denny, at Cambridge; April 20th., by Markwick; May 15th., by M. Harris; May 24th., 1822, by Kirby; June 6th., by J. C. Dale, Esq., and one so late as September 1st. in 1809. I had one out of chrysalis in my own case, on the 20th. of October, 1860.

The caterpillar occurs from June to September. It feeds on various umbelliferous plants, particularly on the marsh parsley, (*Selinum palustre,*) the wild carrot, (*Daucus carota,*) and the fennel, (*Anethum fœniculum.*)

The Swallow-tail measures, in different specimens, from three inches to three inches and three quarters in the expanse of the wings. The ground-colour is yellow, with black markings. The fore wings have a large patch of black, dotted with yellow at the base, and the front margin is black, with three large black marks. The nerves are also black, as is likewise the hind margin, on which are eight yellow marks, and above is a thick powdering of minute yellow dots. The hind wings are also yellow; the inner margin, and a broad border on the outside, black, the latter with six yellow crescents, above which is a thick sprinkling of blue dots. Near the inside corner is a red eye, margined with yellow beneath and blue above, the latter with a black crescent above it.

The under side of the wings is lighter-coloured than the upper, and the black markings are less extended. A narrow black bar supersedes the yellow crescents, above which the dotting of yellow is more thick. The outside black bar of the hind wings is much lighter-coloured, the black being limited to its curved margins, and in the middle of the hind wings are three triangular red spots, and there is another spot of the same in the yellow spot, next to its front edge.

The caterpillar is green, with velvet black rings, dotted alternately with yellowish red.

The crysalis is light green, with yellow on the sides and the back: this is said of the female. "The colour of the male varies from nearly black to a light brownish rufous, having a darker line down each side and bordering the wing-cases; the two prominences on the front of the head, that on the under side of the front of the thorax, and the inner side of the prominences representing the fore legs of the larva, are dark rufous, nearly approaching to black. The wing-cases are slightly tinged with the same colour, having a few black veins originating at the base, and running down towards the anal angle, giving out branches towards the exterior margin along their whole extent. The characters which appear to be common to both, are the shape and the rufous lines down the sides."

The figure is taken from an unusually fine specimen in my own collection, bred in 1851, from a chrysalis received with others from the Rev. George Rudson Read, who had them from Cambridgeshire by the "Penny Post."

SCARCE SWALLOW-TAIL.

PLATE II.

Papilio podalirius,	LINNÆUS. DONOVAN. CURTIS.
" "	LEWIN. WESTWOOD. DUNCAN.
Podalirius Europæus,	SWAINSON.
Iphiclides podalirius,	HUBNER.

SOME there are who dogmatically deny the claim of this species to be a British insect, but the following facts will be sufficient for every unbiassed judgment. Nothing is more certain, as will abundantly appear in the course of the present work, than that some species once common in particular districts, now are never known there, and, 'vice versâ,' that new ones, new to the district, spring up on a sudden, where none had been ever seen before, in the memory, at least, of the "oldest inhabitant."

This Swallow-tail is a native of Europe, Asia Minor, and the northern parts of Africa. It is plentiful near Moscow and Berlin; in fact throughout the whole of our continent.

The following authorities are extant for its admission to a place in our native fauna:—

Berkenhout, in his "Outlines of British Natural History," says that it is "rare in woods;" and Haworth observes that Dr. Berkenhout might probably have had it, as he had heard of his having given a large price for a rare Swallow-tail from Cambridgeshire.

Mr. Rippon says, in 1778? that twenty-five years previously he had taken "two sorts" of Swallow-tails near Beverley, Yorkshire.

Mr. H. Sims was certain that he saw 'Podalirius' on the 24th. of August, 1810, about twelve o'clock, on his way from Norwich to Salhouse. He struck at it with a forceps, but, for want of a better kind of net, was unable to catch it.

My esteemed friend, J. C. Dale, Esq., the well-known entomologist, is also certain that he saw one settled on some rushes near Eltisley, Cambridgeshire, in July, 1818. The wings were half-expanded towards the sun.

Mr. B. Standish was also certain that he saw this butterfly on or about September 20th., 1829, near Richmond Park.

A friend of his, who was in his company when he saw this one, saw another in 1820.

Dr. Abbot told Haworth that he had seen 'Podalirius' two or three times, previous to his capture of it, presently to be stated.

Mr. Thomas Allis says as follows in "The Naturalist," old series, vol. i., pages 38-9:—"Having noticed a good deal of dissension respecting the genuineness of 'Papilio Podalirius' as a British insect, I take this opportunity of announcing, through the medium of your journal, that I myself possess a pair which I believe to be British. I met with them under the following circumstances:—Happening to be at Portsmouth the summer before last, for the first time, I enquired, as is my usual practice on going to a town before unvisited by me, for collectors of Natural History specimens: I soon found one, and among the collection was a pair of the above-named species. The owner assured me they were British, that they were caught by a person she employed in the neighbourhood, and that she set them up herself. As it would not have been worth her while to have imposed on me in this instance, and especially as she did not seem aware of the value of the specimens, I feel no doubt but that they were really British. She could not at the time exactly inform me where they were taken, but on my return to Portsmouth about a fortnight afterwards, she told me she had learned, from the captor, that they were obtained in the New Forest. From what I have said, I feel justified in considering myself the fortunate possessor of specimens of British Papilio Podalirius."

The above relate only to "ocular demonstration;" now then for those "stubborn things"—"facts." First, I have myself seen, in the cabinet of my friend, the Rev. George Rudston Read, Rector of Sutton-upon-Derwent, near Pocklington, the original specimen which was captured by his brother, William Henry Rudston Read, Esq., of Hayton and York, when at school at Eton. He took it on the wing between Slough and Datchet, Berkshire, before the month of July, about the year 1826. It is a very dark individual.

Again, the late Rev. F. W. Hope captured one in Shropshire in 1822, and saw another on the wing.

Mr. Plymley found the larva near the spot where the Rev. F. W. Hope took the perfect insect, but unfortunately the devouring Ichneumon had made a lodgment, so that it came to nothing. Mr. Plymley had the larva brought to him also more than once, and the perfect insect in 1807, from the neighbourhood of Netley, Shropshire.

There was a specimen in Mr. Swainson's cabinet, which he told Donovan was taken by his brother-in-law, Captain Bray: he believed in the Isle of Wight.

Dr. Abbot took one in the month of May, in Bedfordshire. This specimen is now in Mr. Dale's cabinet, 'in perpetuam rei memoriam.'

The butterfly appears in May and June.

The caterpillar feeds on the apple, sloe, plum, peach, and almond.

The perfect insect measures, in different specimens, from three inches to four inches across the wings. The ground-colour is very light cream yellow. The fore wings have two black attenuating streaks near the body, which meet an apparent extension of them on the lower wings; next to these is a very short one, extending only half-way into the wing; this is succeeded by another long one, which reaches quite across the wing, met by a sort of shadow of it on the hind wing, and this again by another short one, extending only one-third across the wing. Again, there is another long one, reaching nearly across the wing, and then lastly, another long one, which completely borders its outside, edged with a narrow yellow line, of the ground-colour of the wing. The hind wings have a black border following the crescent-shaped undulations of their outside edge, and divided by four or five streaks of blue of the same shape. At the inside corner of the wing is a black spot, with a blue patch of irregular shape in its centre, and bordered above with red, forming an eye. The wings have long tails; their tips are yellow.

The under side is paler than the upper, and the black markings less extended. The band on the middle of the hind wings is composed of two narrow black lines, the outer one of which is edged on the inner side with orange.

The caterpillar is short and thick, especially in the middle, and most so towards the head; narrower towards the tail. It is of a green colour, darkest on the back, and spotted with black, varying to light yellowish, with a faint tinge of red beneath. It has a narrow yellowish stripe along the back, and another along the side, near the legs. On the sides are oblique yellowish lines, dotted with reddish. The head is small.

The figure is taken from the original specimen mentioned above, now in the cabinet of the Rev. George Rudston Read, of the Rectory, Sutton-upon-Derwent, near Pocklington.

3

BRIMSTONE.

PLATE III.

Gonepteryx rhamni,	LEACH. STEPHENS. CURTIS. DUNCAN.
Papilio rhamni,	LINNÆUS. DONOVAN. LEWIN. ALBIN.
Goniapteryx rhamni,	WESTWOOD.
Anteos rhamni,	HUBNER.
Ganoris rhamni,	DALMAN.
Rhodocera rhamni,	BOISDUVAL.

IF the imagination chooses so to please itself, it may look upon our own country as a sort of epitome of the northern hemisphere of the world, as to its natural productions, and thus we shall find that on the south coast, from Torquay to Hastings are our tropics; in the midland counties our temperate zone; in Scotland our Arctic regions; and John o'Groat's House will answer to the North Pole. Correspondingly hereto are our Butterflies localized. Excepting in the case of some chance wanderer, driven by we know not what storm, or tempest, or, "favouring gale," the races are, for the most part, distinct, and those which flourish in the one district, would perish at once in the other, through the difference of climate.

This is a most beautifully-coloured butterfly, whether seen on the wing, in its zigzag, impetuous, and hurried flight, or examined at leisure in the cabinet. In the latter case you seem never to be able to get it in a sufficiently good light, so full and bright are its colours, to admire it as it ought to be admired, and when first emerged from the chrysalis, and glittering before you in the sun, it is indeed a most attractive object.

In Yorkshire, and as far north as Newcastle in Northumberland, and the lake district in Westmoreland, it occurs, but not by any means so plentifully as in the more southern parts of the country. It is very rare in the neighbourhood of Falmouth: W. P. Cocks, Esq. has, however, taken it there in 1845, and also in 1850. In Scotland it appears to be unknown.

There are two broods, one in May, the other in the autumn. Many of the autumnal brood live through the winter, and are to be seen in the spring even so early sometimes as February and March, called forth from their retreats by the heat of the returning sun.

The Brimstone is a very discursive insect, and is to be found in gardens, lanes, and fields, especially in those in which clover is grown.

The caterpillar feeds on the buckthorn, *(Rhamnus catharticus;)* the berry-bearing alder, *(Rhamnus frangulus;)* and the broad-leaved buckthorn, *(Rhamnus alaternus.)*

The expanse of the wings in this species varies from rather more than two inches to three inches and a half. In the male the whole of the upper wings is a splendid sulphur yellow colour, with an orange spot above the centre of the fore wings, and a larger one similarly on the hinder wings. A line of the same colour, enlarged here and there into a minute dot, borders the upper corners of the fore wings; underneath, the colour is much fainter, with a cast of green in it: the spot is replaced by a ferruginous dot, whitish in the centre, between which and the margin spoken of is a row of brownish dots.

The female is much paler in colour, resembling more that of the under side of the male.

The caterpillar is green, dotted or irrorated on the back with black; there is a pale green or whitish line on each side, shading off on the upper edge into the green of the rest of the body.

The chrysalis is green, with several reddish dots. It is thickest in the middle, tapering off in front. It is suspended by the tail in an upright position, and retained by a silken thread round the middle of the body. Found July 10th. The chrysalis state lasts about a fortnight, for example, one from July 26th. to August 8th.

A variety, described as a separate species, by the name of "Gonepteryx Cleopatra,' has the upper wings more or less variegated with orange. One of these was taken by John Fullerton, Esq., at Thrybergh Park, near Rotherham, June 27th., 1860.

The engraving is from specimens in my own collection.

 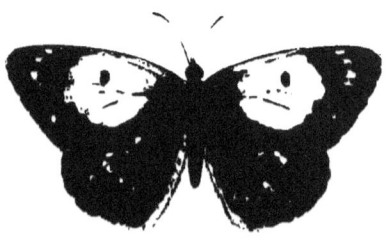

CLOUDED YELLOW.

PLATE IV.

Colias Edusa,	STEPHENS. CURTIS. DUNCAN.
Colias Chrysothome,	STEPHENS.
Papilio Edusa,	FABRICIUS.
Papilio Hyale,	ESPER. DONOVAN.
Papilio Electra,	LEWIN. LINNÆUS.
Papilio Helice,	HUBNER. HAWORTH.

THIS is one of the favourite butterflies of every Entomologist in this country. It is always a valued capture, even though it is met with sometimes in tolerable plenty. It is a fast flyer, and many a rugged chase, when a boy, have I had after it. Some have considered that its appearance, at least in any plenty, is triennial, others quadriennial, and others septennial; but this is not the case, though, certainly, particular seasons are more or less favourable to its development.

Clover fields are a much frequented resort of this beautiful insect, which glitters in the butterfly-collector's eyes as a golden meteor. So also are the grassy cliffs of the sea-shore in those localities where it occurs—these are the "California" of the entomological speculator, and in these he gladly invests his time and trouble.

This species occurs in considerable numbers in those seasons in which it appears, in some of the following, and doubtless in many other localities:—Near Swanage, Lyme Regis, and the cliffs near Charmouth, Dorsetshire, where I have frequently captured it myself in plenty, and where it is to be met with every year, though in some years in greater abundance than in others. Near Worcester, where my brother, Beverley R. Morris, Esq., captured one in 1825; near Broadway, Charing, Feversham, Ramsgate, Margate, Folkestone, Blackheath, and Canterbury, Kent; Broomsgrove, Worcestershire; Dawlish and Exmouth, Devonshire; Biggin, Northamptonshire; and has been taken, as the Hon. T. L. Powys has informed me, in the gardens of Holland House, Kensington, London. Stoke-by-Nayland and Ipswich, Suffolk, as R. B. Postans,

Esq., has informed me; Tiptree, Essex, near the at present celebrated "Model Farm;" Sawbridgeworth, Hertfordshire; once at Leominster, Herefordshire; also near Everton, Nottinghamshire; Weston-super-Mare, in Somerset; Oakley Wood, near Northampton, by the Rev. D. T. Knight; and Swinhope, Lincolnshire, by the Rev. R. P. Alington.

In Yorkshire, in Heslington fields near York, a few in 1833 and 1834, and forty-seven, thirty-seven males and ten females, in 1842; Sand Hutton, and Sutton-on-Derwent, near York, by the Rev. George Rudston Read. One at Liverpool; one at New Brighton, Cheshire; others at Bushmead Priory, Bedfordshire; Lympstone, Teignmouth, Budleigh Salterton, Devonport, and Plymouth, a few, and between Sidmouth and Lyme Regis, in Devonshire, where Thomas Lighton, Esq., saw them in thousands in September, 1843; Broomfield and Chelmsford, Essex; Hitchin, Hertfordshire; near Ely, Reach, Wisbeach, but very rarely indeed, Newmarket Heath, Triplow, Cambridge, Whittlesea Mere, and Horseheath, Cambridgeshire; Lavenham, Sudbury, Foxearth, Great Cornard, Clare, and Kedington, Suffolk; Cromer, Yarmouth, and Roydon, in Norfolk; near Lyndhurst, Winchester, Alverstoke, Christchurch, Southampton, and in the New Forest, Hampshire; Black Park and Chenies, Buckinghamshire; Kemp Town, Brighton, Lewes, Casham, Chichester, Rottingdean, and Arundel, Sussex; Barham Downs, Darenth Wood, and Victoria Park, Hackney; Dorchester and Portland, Dorsetshire; in Shropshire; near Leicester, Shardlow, and Market Harborough, Leicestershire; Headley Lane, Wandsworth, Camberwell, Riddlesdown, Fetcham Downs, Godalming, and Wimbledon, in Surrey. Looe, in Cornwall; Higham and Forest Hill, and the Isle of Wight. One was taken at Appleby, Westmoreland, on the 11th. of September, 185(?), by Henry Moses, Esq., M.D., which is very far north for it, and a second on the 15th., "par nobile." In Northumberland one was seen at Norham, near the Tweed, on the 12th. of September, 1859, by Mr. William Moore. In Wales near Llandudno.

In Scotland only one has been recorded; it was taken by Wyville T. C. Thompson, Esq., on a steep bank near the sea, in the neighbourhood of Lamlash, in the Isle of Arran.

In Ireland two were seen, and one of them captured by Mr. Joseph Poole, of Grovetown, near Wexford, in that locality, on the 9th. of September, 1844.

It will be observed, from the foregoing accounts, that the southern and south-eastern coasts are by far the most productive localities for this beautiful insect, very few indeed being elsewhere met with.

About the last day of August, or the "First of September," the elegant Clouded Yellow usually makes it appearance. It has however

CLOUDED YELLOW.

been frequently noticed in the beginning of the former month, and even at the end of July. Mr. J. F. Stevens also took one on the 23rd. of that month. One was observed by Mr. S. Stevens on the 29th. of June, 1851, near Higham. It was a fine fresh specimen, and it is the earliest record of its appearance that I have ever heard of, excepting the 16th., 18th., and 24th. of that month, in 1831. J. C. Dale, Esq. has captured it on the 11th. of July, 1811; the 14th., 1818; 18th., 1832; 23rd., 1822; 25th., 1826; 28th., 1818; and 30th., 1808. Alfred Greenwood, Esq. has taken it on the 10th. of July. William Arnold Bromfield, Esq. noticed it in the Isle of Wight, in 1845, from July the 3rd. to October the 29th. Frederick Bond, Esq. has seen it on the 14th. of July, and heard of another taken the same day; the Rev. Edward Horton on the 5th. of August; R. C. R. Jordan, Esq. caught one in fine condition on the 4th. of November, 1843; and J. C. Dale, Esq. one on the same day of the same month, in 1808. The late Captain Blomer took one on the 3rd. of November.

The caterpillar feeds on the *Medicago lupulina*, various species of clover (*trifolium*), etc.

The male Clouded Yellow measures from two inches to two inches and a half across the wings. The fore wings are of an exceedingly rich and lovely orange-colour, with a rounded black spot near the centre, and a broad black margin irregularly indented on the inner side, with several narrow orange nerve-like lines running across it. There is also an elegant very narrow pink and light orange border outside the black border, at the extreme edge. The hind wings are of a deeper orange-colour, with a large round central spot of a brighter and very beautiful hue, darkened at the edge.

The female has the broad black border on the fore wings interspersed with several irregular yellow marks, as if the ground-colour of the wings shewed through. The hind wings are darker, and of a yellower tint than in the male, with a shade of green; and their black margin is singularly interrupted with yellow. Underneath, the fore wings are of a lighter orange-colour, with a black central spot; the margins greenish, with a row of blackish spots at some distance from the lower part of the outside margin. The hind wings are greenish orange, with a round dull silvery spot, surrounded with red, and attended, in some specimens, by a satellite smaller silvery dot. Between it and the outside margin is a row of reddish brown dots; one of them large, in the direction of the middle of the upper side, the others very faint.

There is a permanent variety of the female of this species—the 'Colias Helice' of some Authors, which is occasionally, though but rarely, met with. It is a very interesting insect. The ground-colour of the wings is pale yellowish white, as are the light spots on the outside margins of both wings. One has been taken with a tinge of orange.

The white variety has been taken near Charmouth, by my friend, Henry Arthur Beaumont, Esq., and seen by Thomas Lighton, Esq. between Sidmouth, Devonshire, and Lyme Regis, Dorsetshire, as also at Glanville's Wootton, in the latter county; near Teignmouth, Devonshire, by R. C. R. Jordan, Esq.; Brighton, Sussex; also in the Isle of Wight. One at Carisbrooke Castle by the late Captain Blomer.

In a variety in my own cabinet, the only one of the kind that I have seen, there is a divided black streak connecting the central black spot on the fore wings with their black side border.

Another variety in this country is of very small size, and has been erroneously made a separate species, as 'Colias Chrysothome.'

One was seen by the Rev. J. F. Dawson, in which one of the fore wings was white, and all the other three orange-colour.

One in my collection, captured by my brother, Frederick Philipse Morris, Esq., near Charmouth, Dorsetshire, is the largest that I have ever seen. Its wings expand to the width of nearly two inches and three quarters.

The figures are taken from specimens in my own collection. One of them the unusually large one just referred to.

PALE CLOUDED YELLOW.

PLATE V.

Colias Hyale,	OCHSENHEIMER. LEACH. STEPHENS.
" "	CURTIS. DUNCAN.
Papilio Hyale,	DONOVAN.
Papilio Palæno,	ESPER.

This beautiful butterfly is proverbial for the uncertainty of its shewing itself. One year many will be taken in various parts of the country; the next, scarce one will be seen. The proper time of its first appearance is the last week in August, but sometimes it is later than the preceding species.

Clover fields, tre-foil, saint-foin, and lucerne fields, sunny grass banks, and various other situations, are its resort.

The southern districts are obviously the "locale" of this species.

It is plentiful in Africa, in the northern parts of Asia, in Nepaul, Cachemere, and other countries, and also in Europe.

It has been captured or seen in Heslington fields near York, in 1842; Winchester in Hampshire; near Dover, Birch Wood, Darenth Wood, Margate, Charing, and Headley Lane, Kent; Lyme Regis, Dorsetshire; Wolverton; the Isle of Wight; Lewes, both on the Downs and the Ringmer Road, near Shoreham, Kemp-Town, and Brighton, Arundel and Worthing, in Sussex; Matlock, Derbyshire; Eton, Buckinghamshire; Lincoln; near Cambridge, Wisbeach, but very rarely indeed, and Triplow, Cambridgeshire; near Leicester; in Northamptonshire; Broomfield, Dedham, Colchester, and Epping, in Essex; Shelly, Stoke-by-Nayland, and Ipswich, Suffolk; Belton, Norfolk; Exmouth, Devonshire; at Falmouth, in Cornwall, as W. P. Cocks, Esq., has informed me; and several other localities. In Wales near Llandudno.

It is said to be double brooded, appearing in May, and in August or September.—August 14th., 21st.

The Pale Clouded Yellow measures from two inches to two inches and a quarter in the expanse of its wings. In the male the ground-colour is sulphur yellow, with a black central spot, and an irregular broad band, edged with light pink on the outside margin, in which is an interrupted series of spots of the same hue as the ground-colour of the wings. The hind wings have a light orange-coloured central spot, and the margin at the corner is partially and irregularly bordered with black, edged with light pink.

In the female the ground-colour inclines more to very light cream-colour: underneath, the fore wings are pale yellow, the outside corner orange yellow; a row of blackish transverse marks runs parallel to, but at some distance from, the outside margin. The central spot is black, yellow in the centre. The hind wings are orange yellow, with a large silvery spot, accompanied by a small eye-like dot, surrounded with reddish. There is a row of small blackish spots between these and the hind corner, round the edge of the wing.

The caterpillar is of a velvet green colour, with two stripes of yellow on the sides, and black dots on the segments.

The chrysalis is green, with a yellow line on the sides.

Varieties of this insect have occurred: one is described as of a whitish colour; another with the rich sulphur band that divides the broad black margin at the outside corner of the fore wings, uninterrupted, and the central spot of intense blackness; the hinds wings more than ordinarily rounded at the margins, very faintly marked with black, and the central spot or spots scarcely discernible: the size of the insect smaller than usual, and its whole contour different. Another was seen which, as well as could be observed, was of a very rich sulphur-colour, and the outside corner margins, and central spot of a rather deep red.

The figure of the female is taken from a specimen in my own cabinet.

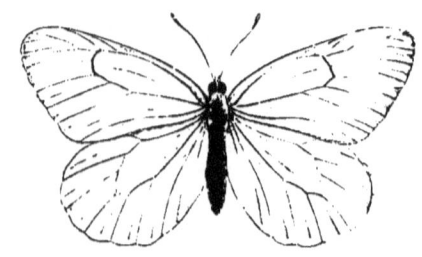

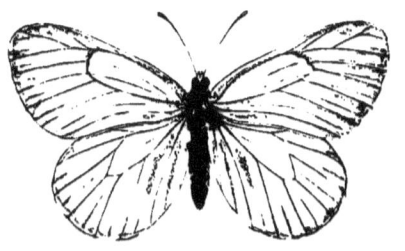

BLACK VEINED.

PLATE VI.

Pieris cratægi,	SCHRANK. LATREILLE. BOISDUVAL.
" "	STEPHENS. CURTIS. DUNCAN.
Papilio cratægi,	LINNÆUS. LEWIN. DONOVAN.
" "	ALBIN. WILKES.
Pontia cratægi,	FABRICIUS.
Luconea cratægi,	DONZEL.
Aporia cratægi,	HUBNER.

THE remark made in a previous article as to an imaginary hemisphere, may be carried still farther by confining it to each one's separate county; thus, the warm sandy soil in the extreme south of Yorkshire—in the Doncaster neighbourhood, will be found to be rich in insect life; the mountains of Craven to have their Alpine productions; flat Holderness those which are attached to a low situation; and "The York and Ainsty" entomological hunters will find their game in the coverts that protect it there.

On the continent this butterfly is so very common, and occurs in some seasons in such prodigious numbers as to cause serious damage, in the caterpillar state, to gardens.

The Black-veined White appears the end of June and beginning of July.

This species, a very local one, is plentiful near Feversham, in Kent, where my friend, the Rev. Henry Hilton, has taken it in former years, the Blean Woods, near Canterbury, and Knockwood, near Tenterden; on the hill side near Cracombe House, Evesham, Worcestershire, where my friend, Hugh Edwin Strickland, Esq., when he resided there, used to see it in abundance; Sywell Wood, near Northampton; and Barnwell and Ashton Wold, and the neighbourhood of Polebrook, Northamptonshire, as the Hon. Thomas Littleton Powys has informed me. Lyndhurst and the New Forest, in Hampshire; and Combe Wood, Surrey; between Stilton and Alconbury, Huntingdonshire; Enborne Copse, near Newbury, Berkshire, the residence of the famous "Jack of Newbury;" Chelsea, Middlesex; Muswell Hill and Herne Bay, Kent; Lewes, in Sussex; and Glanville's Wootton, near Sherborne, Dorsetshire, are also given as localities for it; but I believe it is no longer found in the last-named situation. It has been taken, I am told, at Bishop's Wood, Cawood, Yorkshire.

The caterpillar feeds on the whitethorn, (*Crataegus oxyacantha*,) the blackthorn, (*Crataegus nigra*,) the cherry pear, the *Prunus spinosa*, and other fruit trees.

This butterfly varies in size from about two inches and a quarter to nearly three inches: all the four wings are of a dull milk-white colour, elegantly, at least in the eye of the entomologist, streaked over with the black veins from whence the insect derives its name; they shew through, the wings being semi-transparent, so that the under side resembles the upper in its markings and general appearance.

In the female the veins of the fore wings are generally of a brownish hue; and in one specimen that I have, the outer edge is bordered with a very deeply indented line of blackish brown, the indentations running up the veins to a point, but all united together at the outside.

The caterpillar is at first black, but becomes afterwards thickly covered with whitish hairs, and on the sides and underneath is of a dark grey colour, with two longitudinal stripes of red or yellow.

The chrysalis is greenish white, with two streaks of yellow on the sides, a number of black dots, and a few black streaks.

The figures are taken from specimens in my own collection.

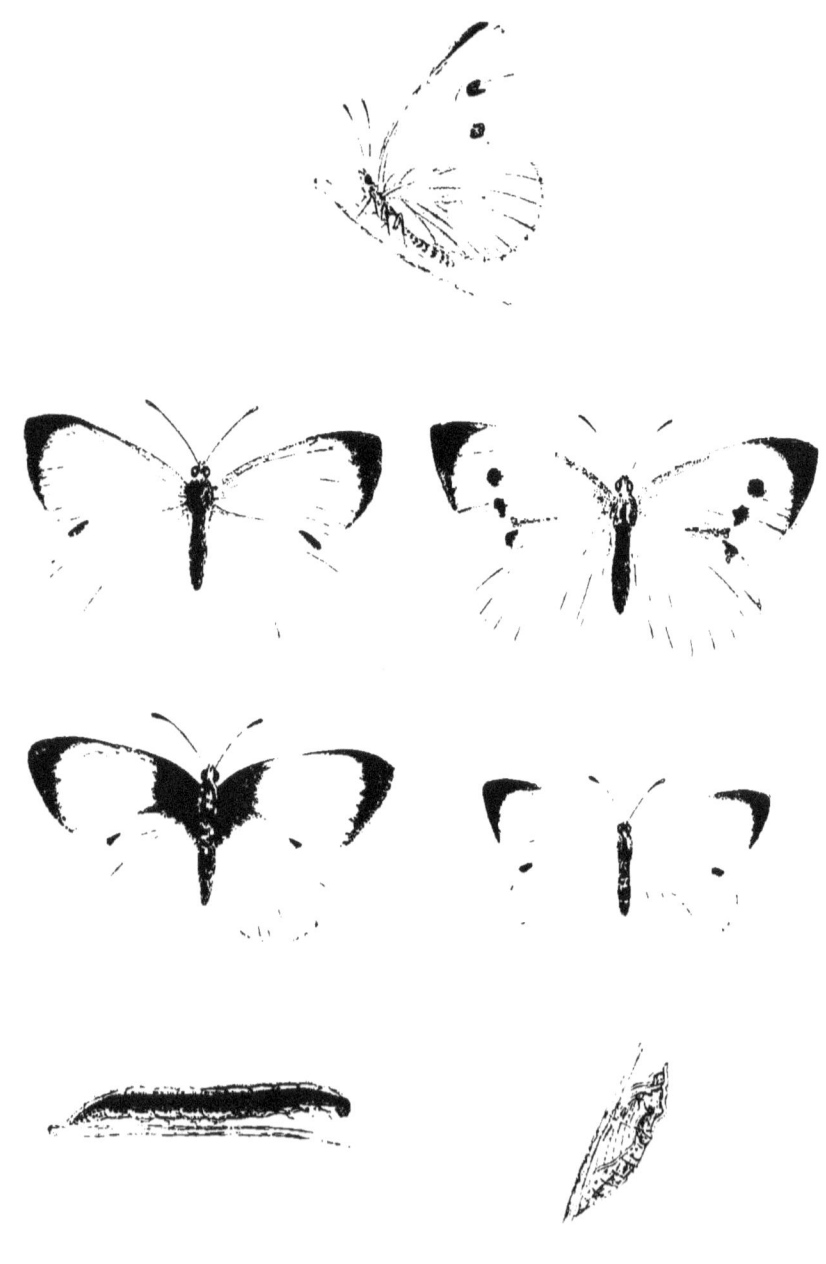

LARGE WHITE.

PLATE VII.

Pontia brassicæ,	FABRICIUS. OCHSENHEIMER.
" "	LEACH. CURTIS. STEPHENS. JERMYN.
Calophaga brassicæ,	HUBNER.
Papilio brassicæ,	LINNÆUS. DONOVAN. LEWIN. HAWORTH.
Ganoris brassicæ,	DALMAN.
Picris brassicæ,	SCHRANK. LATREILLE. BOISDUVAL.
" "	ZETTERSTEDT.

THE Large White is very common throughout Europe, and also, according to some authors, in the north-east of Africa, and even in Eastern Asia and Japan.

It is a very abundant species in this country, and its caterpillar causes much damage in gardens in dry seasons which are favourable to their production.

The perfect insect occurs about the middle of May or earlier, if what we may now call the "Queen's Own" days shine upon it in its hidden existence. One was seen near Doncaster, on the 3rd. of February, 1837, and that during a severe frost. A second brood appears in July and August.

The eggs of the first brood are laid about the end of May, and the caterpillars are hatched the beginning of June. They turn into the chrysalis state at the end of that month, and the fly emerged in about a week or a fortnight, according as the season is less or more favourable to its development. The eggs of the second brood produce caterpillars which turn into chrysalides in the course of the autumn, and remain in that state until the following May.

The caterpillar feeds on the common cabbage, (*Brassica oleracea*.) The caterpillar is found in July and October.

The expanse of the wings varies ordinarily in different individuals, from two inches and a half to two and three quarters—a singularly small specimen in my own collection, figured in the plate, measures

D

under two inches across. The upper surface of all the wings is white, the fore ones having a broad black patch following the angle of the outside corner, and indented on its lower inner edge. There is a pale blackish grey margin at the front of the wings, extending nearly to the patch at the corner; and the whole front edge has a line of black along it. Occasionally, but very rarely, the males have a black spot on the fore wings; the hind wings have a black spot on the middle of their front edge. Underneath, there are two black spots on the fore wings, which however are independent of any generally apparent corresponding markings on the upper side. The hind wings are dull yellow, minutely dotted all over with black specks. There is only a faint trace of the black mark on the middle of the front edge.

The female has the outside patch larger than in the male, with its inner edge more or less deeply indented. There are also two large irregular round black spots on the middle of the fore wings, and immediately beneath them a streak of the same running along the lower edge of the wing, and attenuating towards the body. The hind wings have a triangular-shaped spot on their front side in a line with the spots on the fore wings. Underneath, the fore wings shew the black spots plainly through, and the black patch very faintly. The hind wings are of a dull greenish colour, dotted all over with very minute black spots, and shewing the black triangular patch rather obscurely through.

The eggs are deposited in clusters.

The caterpillar is greenish yellow, the segments being almost covered with black tubercles of different sizes, from each side of which arise white hairs, three of the larger ones forming a triangle. The head, fore legs, and hind segment are also black. There is a line of green down each side, and one along the back.

The chrysalis is pale green, spotted with black, and with three yellow lines.

Varieties occur both in size and markings. A very remarkable one, figured in the "Zoologist," page 471, is given in the plate. It was taken in a garden in Leicester, in the year 1843.

Some have imagined a separate species under the name of 'Pontia Chariclea.'

The engravings are from specimens in my own collection; one of them the unusually small one before referred to.

8

SMALL WHITE.

PLATE VIII.

Pontia rapæ,	OCHSENHEIMER. STEPHENS.
" "	CURTIS. DUNCAN.
Papilio rapæ,	LINNÆUS. HAWORTH.
" "	LEWIN. WILKES.
Pieris rapæ,	LATREILLE. BOISDUVAL. ZETTERSTEDT.
Ganoris rapæ,	DALMAN.
Catophaga rapæ,	HUBNER.

THIS common species inhabits the whole of Europe from north to south, and is found in various parts of Asia, and the north of Africa.

An extraordinary migration of this butterfly from France to Dover was witnessed on the 5th. of July, 1846; and the "Canterbury Journal" recorded at the time that such was the density and extent of the cloud formed by the living mass, that it completely obscured the sun from the people on board the continental steamers on their passage, for many hundreds of yards, while the insects strewed the deep in all directions. The flight reached England about twelve o'clock at noon, and dispersed themselves inland and along shore, darkening the air as they went. During the sea passage of the butterflies, the weather was calm and sunny, with scarce a puff of wind stirring, but an hour or so after they reached 'terra firma,' it began to blow "great guns" from the S.W., the direction whence the insects came. The gardens suffered from the ravages of their larvæ, even at the distance of ten miles from Dover.

In this country the Small White is very abundant, and there are three broods, one appearing towards the end of April, and the other about the beginning of July, followed by a third in September.

The date of the appearance of the caterpillar is in July and August. It feeds on the cabbage, *(Brassica oleracea.)* It remains in chrysalis about a fortnight, more or less.

The expanse of the wings varies from one inch and three quarters to nearly two inches and a half; a singularly small one, captured some years since, by my brother, Frederick Philipse Morris, Esq., is figured in the plate. The colour of this insect is milk-white; the fore wings have a dusky or black mark, irregularly defined at the tip, extending along part of the margin; and there is a black spot near the middle of the wing, and a second indistinctly visible. The hind wings have a dull dusky black mark in a line with them, about the middle of the fore edge. Underneath, the mark at the tip shows through, of a pale yellowish colour, and there are two black spots near the centre; in fact the same as those, one of which only is apparently visible, on the upper surface. The hind wings are yellowish, thickly irrorated, principally near the base, with minute dots.

The female has two black spots near the centre of the upper side of the fore wings, and in many instances an elongated patch of dusky black on the lower margin. There is a dull black mark toward the centre of the fore margin of the hind wings, in a line with the two on the upper wings. Sometimes the whole upper surface is of a pale buff or yellowish colour.

The eggs are placed singly.

The caterpillar is pale green, with a narrow line of orange yellow along the back, and an interrupted line of yellow on the lower part of each side. The head, feet, and tail are entirely green. The body is slightly striated across with the segments.

This is a most variable insect, especially in the size of the spots on the upper wings; in some, in fact, they are wholly obliterated, and in others they are very large. I have a whole row in my cabinet, no two of which are exactly alike.

The figures are taken from some of them—one the singularly small one already spoken of.

J. C. Dale, Esq. has two which seem intermediate between this and the following species.

GREEN VEINED.

PLATE IX.

Pontia napi,	FABRICIUS. OCHSENHEIMER. CURTIS.
" "	STEPHENS. DUNCAN.
Papilio napi,	LINNÆUS. LEWIN. DONOVAN.
" "	ALBIN. WILKES.
Pieris napi,	SCHRANK. LATREILLE.
" "	BOISDUVAL. ZETTERSTEDT.
Ganoris napi,	DALMAN.
Catophaga napi,	HUBNER.

COMPARATIVELY plain as this insect is, yet, looking at it, as at all others, with the eyes that the entomologist does, he will always say "Who can paint like nature?"

The Green-veined White is another of our most common native species. It occurs about the middle of May, and also in July, and is found in all situations—gardens, woods, lanes, and fields.

The caterpillar feeds on different species of *Brassica*, *Reseda*, *Raphanus*, and other plants.

This species varies greatly in size, some being only about an inch and a half in width, and others as much as two. A very small one, captured I believe by myself some years since, and figured in the plate, is only an inch and a quarter across the wings.

The wings are white, dusky black at the tips and the base; and there is generally a black spot not far from the outside edge of the fore wings. Some however have no spot whatever, or the very faintest indication of one, which is more visible if held up against the light. There are some small irregularly-shaped triangular marks at the end of the nerves at the outside edge of the wing: the hind wings are white. Underneath, the fore wings have two spots, as in the female on the upper surface, and the nerves obscure black. The hind wings are pale yellowish, with the nerves broadly margined on each side with dusky greenish, widest on the inner part, and tapering off afterwards to the edge.

The female is generally smaller than the male, and the wings are more rounded. They are of a light greenish white colour, veined with dusky black; the tips dusky black, and there are two spots of the same towards the outer margin of the wing; the lower one, at the lower edge, running into a wide streak, which runs up to the base of the wing. The hind wings are also streaked with dusky, but more faintly, and there is one spot on their upper edge, in a line with the two on the upper wings: underneath, the streaked nerves shew through, as also the spots. The hind wings are pale yellowish, the nerves streaked with dull greenish on both sides, widest above, and each running off to a point at the outer edge.

The caterpillar is of a dull green colour on the back, the sides brighter, with red dots placed on yellow spots on each segments.

The chrysalis is greyish or yellowish green with black spots.

This is another very variable insect. One has been described as a separate species, under the name of 'Papilio napeæ' and 'Pontia napeæ.' The male has the whole upper surface of the fore wings white, with the tip, a spot, and two or three triangular-shaped markings on the hind margin, black; the hind wings white, with the nerves near the base widened and greenish. Underneath, the fore wings have the nerves rather widened into greenish streaks, with two ash-coloured spots placed transversely, and the tips yellowish; the hind wings pale yellowish with one deeper streak.

The female has the fore wings, with the tips, and three spots, one of which is nearly triangular, dusky black; the hind wings clearer yellow. Underneath, the hind wings have the streaks on each side of the nerves more or less wide in different specimens.

Another variety, also erroneously made into a species, under the name of 'Pontia sabellicæ,' has the veins strongly margined on each side with brown, and the fore wings of a rounded shape in some specimens, and in others only so on the lower part of the margin, which is therefore widened more than ordinarily.

The male of our present species has been known with the wings of the rounded shape which usually is characteristic of the females.

The figures are taken from specimens in my own collection, one of them the very unusually small one already referred to.

CHEQUERED WHITE.

BATH WHITE. SLIGHT GREENISH HALF-MOURNER.
VERNOUN'S GREENISH HALF-MOURNER.

PLATE X.

Pontia Daplidice,	FABRICIUS. OCHSENHEIMER. CURTIS.
Papilio Daplidice,	LINNÆUS. LEWIN. DONOVAN.
Pieris Daplidice,	SCHRANK. LATREILLE.
" "	BOISDUVAL. ZETTERSTEDT.
Mancipium Daplidice,	STEPHENS. DUNCAN.
Synchloe Daplidice,	HUBNER.

EXCEEDED though this lovely insect is by many of brighter colouring, yet it needs not the additional enhancement of its great rarity to make the collector exclaim, " Can imagination boast, amid her gay creation, hues like these?" This is indeed a prize in his harmless lottery; one which it falls to the lot of but very few to gain. It must be a singularly fortunate day in the year that is not a blank one in regard to the capture of the Chequered White.

The Chequered White, or Bath White, is very common in many of the southern parts of the continent of Europe, as well as on the opposite coasts of Africa, in Barbary, as also in Asia Minor and Cashmere, and no doubt in many other parts of the Asiatic continent. It is mostly found in dry and sandy situations.

In this country, as before pointed out, it is very rare. Ray has recorded that it was formerly taken by Vernon, near Cambridge; and Petiver that it was found near Hampstead. Lewin says that one was taken near Bath—whence one of its names, and that the fact had been chronicled by a young lady in needlework, in which the fly was depicted. Haworth states that one was taken in June in White Wood, near Gamlingay, Cambridgeshire. Stephens captured one on the 14th. of August, 1818, in the meadow behind the Castle of Dover, where others have since been taken by Mr. Le Plaistrier, of that place;

also more near Deal, one at Kingsdown, one near Walmar, and another at Tenterden, in the same county. My excellent friend, J. C. Dale, Esq., has also recorded the capture of one about the same time near Bristol. One was taken at Keynsham, between Bath and Bristol, and one at Whittlesea, in Cambridgeshire, in 1852, by E. Burton, Esq., of Spekelands, near Liverpool. One near Lewes, Brighton, and four near Horndean, Hampshire.

Two broods occur in the year, the earlier one in April and May, and the latter in the end of July and August.

The caterpillar feeds upon wild woad, base-rocket, and wild cabbage, as also on various species of *Reseda* and cruciferous plants.

It is described by Boisduval as being of a bluish ash-colour above, and on the sides covered with small black raised dots, and four white stripes along the sides; beneath, whitish, as are the legs, each with a yellow spot above it.

The chrysalis is greyish, dotted with black, with several reddish stripes.

This fly varies in the expansion of its wings, from not quite an inch and three quarters to nearly two inches. The wings are white, with a shade of cream-colour. The fore wings, which are unusually pointed, the outer margin being slightly concave, are blackish at the base, and there is a rather large black spot about the centre of the wing, where the transverse veins appear of a white colour. The tip is irregularly marked with black, irrorated with white, which is widest towards the front margin; the black patch is also marked with four irregular white spots. The hind wings are white, the markings of the under side shewing faintly through.

Underneath, the marks of the fore wings are of a greenish colour, and there is a spot on the inner edge. The hind wings are yellowish green, with three large white spots, forming a triangle, towards the outer corner of the wing, succeeded by an irregular white bar beyond its middle, crossed by yellowish veins, and with five white club-shaped spots on the outer margin.

In the female the black patch on the fore wings, which are of a convex and rounded form, is darker than in the male, and there is another small black patch near the inner margin. The hind wings are white, but the markings of the under side show through rather more distinctly than in the male, especially along the outer edge. Underneath they are greenish, and marked as in the male.

WOOD WHITE.

PLATE XI.

Leucophasia loti,	RENNIE.
Leucophasia sinapis,	STEPHENS. BOISDUVAL. DUNCAN.
Papilio sinapis,	LINNÆUS. LEWIN.
" "	DONOVAN. HARRIS.
Pontia sinapis,	FABRICIUS. OCHSENHEIMER. LEACH.
Leptoria candida,	WESTWOOD.
Pieris sinapis,	SCHRANK. LATREILLE. GODART.
Ganoris sinapis,	DALMAN.
Leptoria sinapis,	HUBNER.
Papilio candidus,	RETZIUS.

WELL is it for the entomologist that his is "untaxed and undisputed game." He wants no "license" to saunter harmlessly in quest of the trophies of his skill, through the winding lanes of his native country, and the green pathways that labyrinth her woods. Sometimes, indeed, some stupid churl is fain to exert, and probably to overstrain, his deputed authority, but for the most part, the land is as free as the air to the peaceful insect hunter.

The Wood White is a very pretty object, floating lightly in the glades of the wood, in a slow and undulating manner. It appears, according to some accounts, to be double-brooded, the first appearing at the end of May, and the second in August.

I have once taken this interesting insect, in the year 1837, in Sandal Beat, near Doncaster, Yorkshire, and repeatedly in the Ran-Dan Woods, a most excellent locality for many good species, near Broomsgrove, Worcestershire. It occurs also in the following localities: —Barnwell and Ashton Wold, and the neighbourhood of Polebrook, Northamptonshire; near Carlisle, in Cumberland; rarely near Great Bedwyn and Sarum, Wiltshire, as J. W. Lukis, Esq. has informed me; in the woods on the banks of the river Dart, in Devonshire, as James Dalton, Esq., of Worcester College, Oxford, has written me word; Coggeshall and Raydon Woods, near Hadleigh, Essex, as R. M. Postans, Esq. has informed me; Grange, near Ulverstone, in Lancashire; Lewes and Brighton, in Sussex; and very abundantly in all the woods in the neighbourhood of Ardrahan, in the county of Galway, in Ireland, as I have just learned by an obliging communication from A. G. More, Esq., of Trinity College, Cambridge.

The caterpillar feeds on the different species of *Lathyrus, Lotus,* and *Orobus;* the Vetch, *Vicia cracca,* and, according to Fabricius, the *Sinapis,* or wild mustard; but this is now said not to be the case, so that the specific name thence given to it has been altered.

This fragile-looking butterfly measures from one inch and a half to nearly two inches across the wings. It is of a delicate white colour, with a rounded dull black spot at the tip of the fore wings. In some specimens, however, this spot is nearly obliterated, and in others is entirely wanting.

Underneath, the fore wings have the front margin greyish coloured, and the base and tip of these wings very pale yellowish green. The hind wings are tinted very faintly with greenish yellow, with the nerves, and two irregular, and in many instances interrupted, transverse bars, of a greyish ash-colour.

The female resembles the male.

The caterpillar is green, darker near the end, and with a yellow stripe along the sides, above the feet.

The chrysalis is at first of a greenish colour, but afterwards becomes whitish grey, with red dots on the sides and upon the wing cases.

ORANGE TIP.

PLATE XII.

Mancipium Cardamines,	Stephens. Duncan.
Papilio Cardamines,	Linnæus. Lewin. Harris.
" "	Wilkes. Donovan.
Anthocharis Cardamines,	Boisduval. Godart.
Pieris Cardamines,	Schrank. Latreille.
" "	Zetterstedt.
Ganoris Cardamines,	Dalman.
Euchloë Cardamines,	Hubner.

The endless variety of nature is perhaps the most wonderful feature in it. Varied indeed are the combinations which the mind of man strikes out, but still there is a something, a 'je ne sçai quoi,' which, though the gift that enables him to do even what he does, is itself a natural gift, yet leaves it always manifest that his skill in contrivance is but imperfect. The varieties that he sees in natural objects would never have occurred to his own imagination, however fertile it may be. There is but One of whom we can say with truth, "His work is perfect."

The Orange Tip is a strikingly handsome insect, whether seen in its flight, or adorning the cabinet. I have noticed how very suddenly, from whatever cause, all, or nearly all, disappear when their "little day" is over.

It is a very abundant species in all parts of England, among others at Brighton, Anstey, Bisterne, and I have seen it in great plenty in Ireland, in the grounds of Rostellan Castle, the very beautiful seat of the Marquis of Thomond.

The butterfly appears at the end of May; and Mr. Stephens says that he has had some come out of chrysalis in the middle of June, and others in the middle of July. I saw one in the neighbourhood of Broomsgrove on the 4th. of April, 1868. On the continent there is a second brood, and Mr. Dell, of Plymouth, has taken one near there in that month,—I mean in July.

It is found in all sorts of situations, in the green lane, and the open pathway or riding in the wood; in the sunny meadow and the cultivated garden.

The caterpillar feeds on the *Cardamine impatiens, Turritis glabra, Brassica campestris,* and other plants.

The wings expand to the width of from one inch and three quarters to two inches. Their ground-colour is white; the upper wings are black at the base, and have a black mark, varying in shape, near the centre. The whole of the space between this, and indeed from a little inside it to the tip, is a lovely orange colour, bordered on the outside corner with brownish black, irrorated with very minute orange specks, and indented on the inner side. The hind wings are also black at the base. Underneath, the fore wings resemble the upper, except that there is a little dash of very pale yellow near the base, and the dark mark at the tip is exchanged for dull white, barred with dull green. The fore edge has a few small back dots. The hind wings are most elegantly varied with green marks, and yellowish green, the ground colour being white, and some of the veins yellowish.

The female is without the orange tip; in other respects she resembles the male, but the green underneath is darker.

The caterpillar is green, finely dotted with black, and with a white stripe along the side.

The chrysalis, of a pointed shape, is at first green, which in a few days changes to dull light yellowish grey; the stripes being brighter.

This insect varies very much in the extent of the wings, and also in the size and shape of the black spot on the fore wings. Stephens describes one with the black mark on the fore wings almost obliterated, and with a black spot on the upper surface of the hind wings. Haworth mentions one as having the orange mark on the upper surface of the fore wings almost invisible; and Boisduval another, a female, which had an orange spot on the under surface of the fore wings. Mr. Robert Calvert, of Bishop-Auckland, has written me word of one he has, which measures only one inch and a quarter across the wings.

I think it is also more than ordinarily subject to malformation. I have one which has not only the wing, but the antenna on one side smaller than on the other.

MARBLED WHITE.

HALF-MOURNER. MARMORESS.

PLATE XIII.

Hipparchia Galathea,	LEACH. STEPHENS.
" "	CURTIS. DUNCAN.
Papilio Galathea,	LINNÆUS. LEWIN. WILKES.
" "	DONOVAN. HARRIS.
Arge Galathea,	BOISDUVAL. HUBNER.
Satyrus Galathea,	LATREILLE. DUPONCHEL.

THE entomologist may, or may not be, an aristocrat; but whether or no, politics he eschews. The peer and the plebeian 'æquo pulsat pede' the wood side, or the green lane, the mountain top, or the sheltered valley. The butterfly-collector's pride of race is centred in one which is alien to his own. "There is my friend the weaver" says the excellent poet Crabbe, speaking of an entomological one; and the honest artizan or mechanic will be "hail fellow well met" with the "Proud Duke of Somerset" himself, if both should meet together on common ground, in the kindred pursuit of a rare species.

Thus, the term a "good neighbourhood" may be understood in another sense than that which is commonly meant by it, and I, for one, prefer the retired glade of the forest to "Belgrave Square" or "the Dukeries," and the air of the mountain to that of the Court:—

"Give me but these; I ask no more;
With, etc."

We do indeed lament the loss of many of the "old families," and I must say, speaking entomologically, that we treat with considerable contempt many of our modern new ones.

There is hardly a more strikingly beautiful species of butterfly in our country than the Marbled White: the contrast of its black and white markings is exceedingly pleasing.

It is with us very locally, though widely distributed; in Scotland, however, it is not known.

I have taken this insect in plenty at Pinhay Cliff, Devonshire; near Lyme Regis and Lulworth Cove, Dorsetshire; also at Marr, near Doncaster, and on Buttercrambe moor, near Stamford-Bridge, Yorkshire, in which county it has also been taken at Worst Hill, near Pontefract. It is found in abundance in Hartley Wood, near St. Osyth, Essex, and occurs also at Manningtree, in the same county; and near Great Bedwyn and Sarum, Wiltshire, in isolated spots near woods, as J. W. Lukis, Esq., informs me; Preston, in Lancashire; Sywell Wood, near Northampton, and in plenty in the wood of Ashton Wold, near Polebrook, in the same county, where I have taken it, in company with the Rev. William Bree. In Lincolnshire, in a lane between West Rasen and Kingerby Wood. In Hampshire, Winchester and Lyndhurst, and in the Isle of Wight; in Gloucestershire, near Dursley; in Sussex, near Brighton; and in Essex, in Hainault Forest. It is plentiful near Clonmel, in Ireland.

The perfect insect appears in June and July.

The caterpillar feeds on the cat's-tail grass.

The Marble White varies in the expanse of its wings from two inches to nearly two and a quarter. Its colours are a fine yellowish white and black, with which the whole surface of the wings is chequered over, so that one can hardly say whether the white or the black is the ground-colour. There is a large whitish oval spot near the base of each wing, succeeded by four long whitish patches, the two middle ones being nearest to the outside of the wings, and smaller than the others. Between these and the tip are two smaller white spots, and there is a row of white spots near the margin, divided by a black line, which is again succeeded by the white, forming a margin, interrupted by the continuation of the black which had formed the sides of the white spots before their intersection by the black line. The hind wings have a large oval whitish spot near the base, then an irregular black mark, succeeded by a very broad bar of the former colour, then black again, and then a row of white crescents, varying in size, near the outside margin, divided by a black line, as in the fore wings.

Underneath, the markings correspond, but the black colour is much more faint and indistinct. The fore wings have a small black eye, with a white centre, near the tip. The hind wings have five eyes just above the white crescents near the margins, the third from the outer corner not having an eye, and the eye near the inner corner being a double one. The black markings are irrorated with buff.

The female is of considerably larger size than the male, and the under surface of the wings is of a yellow hue.

Varieties occur, both accidental ones, and others that seem to be permanent; some are quite of a cream yellow on the upper side; others milk white.

One taken near Dover, recorded by the Rev. W. T. Bree, in the "Magazine of Natural History," vol. 5, page 335, and which is figured in the plate, is described as having the upper wings nearly black, except a large white spot near the base, and another divided into three by the veins at the lower edge of the middle part of the fore wings. Underneath, all the wings are clouded with black, and almost entirely without the usual tesselated markings.

In some the black is much suffused over the greater portion of the wings.

In others the black colour is changed into yellowish brown.

Another local variety has the black markings on the under side of the hind wings exchanged for a very light bluff, so that the wings appear nearly white, and without the eyes.

The caterpillar is yellowish green, with a darker line down the back and on each side.

The figures, excepting the one of the variety, are from specimens in my own collection.

WOOD ARGUS.

WOOD LADY. SPECKLED WOOD BUTTERFLY.

PLATE XIV.

Hipparchia Ægeria,	FABRICIUS. OCHSENHEIMER. LEACH.
" "	CURTIS. DUNCAN. STEPHENS.
Satyrus Ægeria,	LATREILLE. BOISDUVAL. DUPONCHEL.
Papilio Ægeria,	LINNÆUS. HAWORTH. DONOVAN.
" "	LEWIN. WILKES. HARRIS. SEPP.
Pyrarge Ægeria,	HUBNER.

"When Flora with her fragrant flowers
Bedeckt the earth so trim and gaye,
And Neptune with his daintye showers
Came to present the monthe of May,
King Henry rode to take the ayre,"

and, at that season, he who cannot ride will walk, and if he have a love for entomology, will turn his steps to the lane or the wood. There he will see the Wood Argus, which delights in such situations, and is a common species in all parts of the country, from the extreme north to the extreme south, and from east to west, from Brighton to Llandudno.

The perfect insect appears in April, June, and August, there being several broods, probably three, in the course of the year.

The caterpillar is found in March, May, and June.

It feeds on various grasses, giving a preference to the common couch-grass.

The wings expand to the width of from one inch and a half to two inches; their ground-colour is greenish brown. The fore wings are marked with a number of pale buff patches, of variable size and of irregular shape, ten or eleven in the strongest-marked individuals, the one nearest the outside corner of the wing having a black eye, with a white dot in the centre. The hind wings have a pale buff patch near the centre of the wing, but rather outside it, and a still smaller and faint one on their fore edge: towards the margin they have four buff patches, the fore one with a very small dot, or none at all, the next with a larger one, and the two hinder ones with still larger ones, each with a dot of white in its centre. In some specimens the black prevails over the buff more than in others, leaving only a narrow border of the latter colour.

Underneath, the brown colour of the fore wings is more clouded, the outer corners being much paler, with the eye near the tip shewing through. The hind wings are more varied with waved shades and lines of a darker and a lighter colour, the upper outer corner being paler, and there is a row of five or six white dots, varying in size, near the outer margin, which is darker, and not unfrequently has a tinge of purple, the larger dots being in fact those of the upper side shewing through.

This is a very variable insect, though preserving on the whole a similarity of appearance. The males are generally smaller and darker in colour than the females, the pale spots on the latter being at the same time larger and more numerous than in the former.

The figures are from specimens in my own collection.

WOOD RINGLET.

PLATE XV.

Hipparchia Hyperanthus,	OCHSENHEIMER. LEACH.
" "	STEPHENS. CURTIS. DUNCAN.
Papilio Hyperanthus,	LINNÆUS. LEWIN. DONOVAN.
"	HAWORTH. HARRIS.
Papili Polymeda,	SCOPOLI. HUBNER.
Satyrus Hyperanthus,	BOISDUVAL.
Enodia Hyperanthus,	HUBNER.

MANY of the praises which good old Izaak Walton bestowed on the art of angling, and especially those which refer to the "higher branch" of it, namely to fly-fishing, apply equally well and suitably to the entomologist's peaceful and gentle pastime. "The murmur of the mountain bee" is a good substitute, at least 'pro tempore,' for the murmur of the purling brook; to catch a rare butterfly is, in its way, as great a pleasure as to catch a trout of three pounds weight, or a fresh-run salmon of a dozen, and in each case alike you have the enjoyment of the fragrant "scent of hawthorn flowers," and of all the other charming scents and sights with which the beneficent Creator has strewed and surrounded the path of those who will seek them in the country, where His hand has placed them.

This insect is plentiful throughout the country generally, but in the neighbourhood of Falmouth it is scarce.

It is found in woods and lanes, and places more immediately adjacent to them.

It appears about the end of June, and in July, lasting to the middle of August.

The catepillar feeds on the *Poa annua* and other grasses, about the roots of which it conceals itself during the day.

The wings expand to the width of from one inch and a half to nearly two inches. The whole of the upper surface is very dark rich brown. The fore wings have one or two small eyes, dark brown, edged with light brown, more or less distinct in different specimens, near their hind margin. The hind wings have two or more similar eyes near their margins, one or two of them with white specks in the centre.

Underneath, the ground-colour is a much lighter brown. The eyes on both the fore and hind wings are much larger than above, there being generally three near the outer angle of the upper wings, the rim, eye, and dot being more or less indistinct, and five much more distinct ones on the hind wings, two and three, the latter inside, but following the lower outer margin, and the former, between the base of the wing and its outside corner, near the fore edge.

The caterpillar is of a greyish white colour, with a slender black line along the back, and sometimes, Mr. Westwood says, it is entirely blackish.

This butterfly is an exceedingly variable one. In some specimens the eyes are very large, and connected together, accompanied by smaller satellite ones. In some the eyes are wholly obliterated on the fore wings, and indeed, strictly speaking, on the hind wings too, there being in their place three minute white specks. A very extensive series of varieties may easily be procured. J. C. Dale, Esq. has one with unequal spots on the opposite wings, and Mr. Wailes, of Newcastle, one with no spot whatever, either above or below.

The figures are taken from specimens in my own collection.

 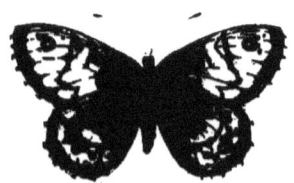

GATE-KEEPER.

SPECKLED WALL. WALL BUTTERFLY.

PLATE XVI.

Hipparchia megæra,	OCHSENHEIMER. LEACH.
" "	STEPHENS. CURTIS.
Papilio megera,	LINNÆUS. LEWIN. DONOVAN.
" "	SEPP. WILKES.
Papilio mægera,	HAWORTH.
Satyrus megæra,	LATREILLE. BOISDUVAL.
" "	DUPONCHEL.
Dira megæra,	HUBNER.
Papilio mæra,	BERKENHOUT. HARRIS.

Those who study what may with truth be called the "Gay Science" of entomology, derive a pleasure which those who devote themselves to mere amusement form and can form no conception of, and that, even from the inspection of those species which present no striking attraction from the beauty of their colours. Such is the fact with respect to the plain insect before us.

This butterfly is to be seen flitting in its zigzag manner along the banks which for the most part it frequents, in July and August. It is fond of settling on walls, whence one of its English names, seeming to take delight in those situations which are the most sheltered, and from which the most heat is reflected. The shadow of your approach disturbs it and you see it flit off, and settle again at some little distance, or continue its irregular flight along the bank.

It occurs from Brighton to Llandudno, and from Bisterne to Doncaster,

The caterpillar is found in the beginning of May, and also in August.

It feeds on different grasses.

The expanse of the wings in this species varies from one inch and a half to nearly two inches. The ground-colour is pale yellowish brown, marked irregularly across with several waved brown bars, and the margins of the same colour; near the tip is a large blackish brown eye, with a white dot, and there is a broad oblique brown bar extending across the middle of the hind part of these wings, which is absent in the female.

The base of the hind wings is brown, and they have, following the outer margin, a row of three or four eyes, varying in size; the middle ones with a white dot in their centre. Underneath, the markings of the fore wings are nearly corresponding to those of the upper side, but the brown bars are not so wide, and the oblique one is wanting. The eye is surrounded by a brown ring, and is accompanied by a smaller satellite.

The hind wings are neatly freckled with brown and ash-colour, with many waved marks of a darker shade, two of them forming a rather broad waved bar across the middle of these wings, beyond which is a row of six or seven eyes, two, three, and two, that at the inner corner being a double one, and these succeeded by a row of darker waves: the eyes are formed of a brown spot, in which is a black dot, and this again has a white spot, more or less distinct, in its centre.

The figures are from specimens in my own cabinet.

ROCK-EYED UNDERWING.

GRAYLING.

PLATE XVII.

Hipparchia Semele,	OCHSENHEIMER. LEACH.
" "	STEPHENS. DUNCAN.
Papilio Semele,	LEWIN. DONOVAN.
" "	HAWORTH. HARRIS.
Satyrus Semele,	LATREILLE. BOISDUVAL. DUPONCHEL.
Eumenis Semele,	HUBNER.

I HAVE seen this interesting insect in the parish of Nafferton, Yorkshire, on the scanty remains of heath on the road to the hamlet of Pockthorpe. The whole of the Yorkshire Wolds, now all enclosed, were, sixty years ago, open downs, with heath and gorse scattered here and there: it is but very little that is left. Then you could ride over the Wolds, so I am told, from Driffield to Malton, twenty miles, without meeting with a hedge or a gate—all was turf—fine old downs: now, 'quantum mutatus,' it is one of the principal corn-growing districts of England. The naturalist cannot but sigh for 'il tempo passa,' with which has flown the Great Bustard, and doubtless many other species of birds once common, or which were at all events far more numerous than they are now, but he must console himself with the absolutely true adage that "whatever is, is best."

The following are also localities, among others, for this species:—York, on the railway near Burnby, at Drewton near Market Weighton, and at Scarborough; Bisterne, in Hampshire; the Isle of Bute and the Isle of Arran; near Durham, South Shields, and Castle Eden Dene, in the county of Durham; the Gog Magog Hills, the neighbourhood of Newmarket and Gamlingay, Cambridgeshire; occasionally near Hunstanton, Norfolk, on some waste herbage, formerly a rabbit-warren; in Wales, near Rhyll and Llandudno; Salisbury Plain, in Wiltshire; Arthur's Seat, near Edinburgh; the Cheddar Cliffs, and Wells, Somersetshire; Walmer Forest and Languard Forest. I have also taken this fly in plenty on the top of the hill between Charmouth and Lyme Regis, Dorsetshire. It is not uncommon near Looe and Falmouth, Cornwall, and is plentiful on Newmarket Heath, in Cambridgeshire, and in various other parts of the county. It occurs sparingly near Great Bedwyn and Sarum, Wiltshire, as J. W. Lukis, Esq. has informed me; also in Hampshire, in chalk-pits, near Winchester, according to J. Wesley, Esq.; and in Sussex near Brighton.

The Rook-eyed Underwing is fond of barren spots, where heath abounds, about stone-pits and rocky places.

The perfect insect appears in the middle of July, and has been known to continue till the 12th. of September.

This butterfly measures from about two inches to two and a half in the expanse of the wings. The fore wings are of a dull brown colour, tinged with bronze, with a broad interrupted bar of various dark patches near the principal vein. Towards the outer margin, are two eyes.

The female is smaller than the male, and the hind wings are brown to the base, with a brighter-coloured wave near the margin, having a single black eye, with a white centre near the inner lower corner. Underneath, the fore wings are darker at the base, with the whole outer part yellowish or pale buff, ended by a narrow dusky margin. There are two eyes, the front one being the larger. The hind wings are marked with numerous narrow white and brown streaks across. The part next the base is the darkest, and is met by a very irregular broad bar of a paler colour, which again becomes darker towards the outside, and near the inner lower corner is a nearly obsolete eyelet, the same indeed that appears also on the upper side.

The caterpillar is green or gray, except on the lower part, which is brownish. There are five longitudinal lines along it, one on the back being darker than the rest.

The figures are from specimens in my own collection.

SMALL MEADOW BROWN.

LARGE HEATH, (GATE-KEEPER.)

PLATE XVIII.

Hipparchia Tithonus,	OCHSENHEIMER. STEPHENS.
" "	CURTIS. DUNCAN. WESTWOOD.
Papilio Tithonus,	LINNÆUS. LEWIN. HARRIS.
Pyronia Tithonus,	HUBNER.
Papilio Tithonius,	VILLARS.
" *Herse,*	HUBNER.
" *Phædra,*	ESPER.
" *Pilosella,*	FABRICIUS. HAWORTH. DONOVAN.

ONE great advantage in the pursuit of entomology is, that no day in the year is, at least no day need be, a 'dies non;' there is always something to be met with, even in the depth of winter, "something to please, and something to instruct." The summer, however, is the hey-day of the butterfly's existence; for though some of them, of several kinds, live throughout the winter in a dormant state, appearing again in early spring, when some hot day calls them forth from their retreat to renewed life, yet it is the former season to which, even to a proverb, they belong.

This is a very common British species, and is widely distributed, occurring in lanes and meadows.

It appears about the middle of July.

The caterpillar is to be found in the beginning of June.

It feeds on the *Poa annua,* or annual meadow grass, and also, according to Haworth, on the *Hieracium pilosella.*

The wings expand to the width of from one inch and a half to an inch and three quarters. Their ground-colour on the upper side is a mixture of red, yellow, and brown. The fore wings are brown at the base and on the front edge, and they have a broad margin also of brown. Near the tip is a large black eye, in which are two small white dots, and a broad brown patch, or rather bar, runs obliquely across the middle of these wings, parallel to the border at the outside edge. The hind wings have a small portion apparent in the centre of the same ground colour as that of the fore ones, but all the remainder is taken up with a dark brown border, and the same colour spread over the base.

Underneath, the fore wings are coloured as above, but the brown patch is wanting. The hind wings are of a golden brown hue at the base and the margin, with an irregular waved greyish buff band running across the middle, with a brown patch near the outer angle, in which are two small eyes, and another patch and eye towards the hind angle, sometimes accompanied by one or two or more small satellite eyes, which vary in size as well as in number.

The female is without the bar which crosses the upper side of the fore wings in the male.

The caterpillar is of a greenish colour, and has a reddish line on each side, and a brownish head.

The figures are from specimens in my own collection.

 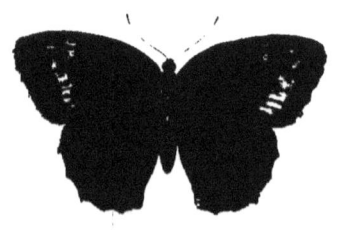

LARGE MEADOW BROWN.

MEADOW BROWN.

PLATE XIX.

Hipparchia Janira,	OCHSENHEIMER. STEPHENS. LEACH.
" "	CURTIS. DUNCAN. WESTWOOD.
Papilio Janira,	LINNÆUS, (*male.*) TURTON. STEWART.
" *Jurtina,*	LINNÆUS, (*female.*) LEWIN. DONOVAN.
" "	HAWORTH. HARRIS.
" *Hyperanthus,*	WILKES. ALBIN.
Epirephile Hyperanthus,	HUBNER.

THIS butterfly is also to be seen in abundance

"In summer time, when leaves grow greene,
And blossoms bedecke the tree."

It is one of our most plentiful species, and occurs in all parts of the country. I well remember the extraordinary numbers in which it appeared in the unusually hot and dry summer of the year 1826.

It appears in June, and lasts till the end of August.

The caterpillar feeds on various kinds of grasses, more especially on the *Poa pratensis*.

This insect varies in the expanse of its wings from one inch and a half to two inches; the whole upper surface of the fore wings is brown, with more or less of a fulvous tinge, with a faint shade of bronze over it. Near the tip is a small black eye with a white dot in its centre, surrounded by a ring of orange buff, sometimes more widely, though indistinctly, extended. In some specimens there are two white dots, and in others more. The hind wings are wholly brown.

Underneath, the fore wings are orange-yellow brown, with a darker border of the same; the eye and eyelet shew through. The hind wings are darker brown, with a broad waved bar at a little distance from the outer margin, and in it from one to three minute dark specks.

The female is larger than the male; the general colours are the same, but the fore wings have a large irregular patch of fulvous brown, around, but chiefly extended beneath, the eye. The general colour of the hind wings is brown, and they have a broad indistinct bar of a paler colour with an orange tinge, following the margin, but a little within it. The brown at the edge of the latter is lighter than the remainder of it.

Underneath, the fore wings are pale fulvous, the eye and eyelet the same as on the upper side, and the large patch shews through from above a little lighter than the surrounding colour, but more distinctly defined by straight edges on its sides. The hind wings are darkish brown at the base with a tinge of bronze. This is succeeded by a paler irregular bar, the same which partially shews through on the upper side, and this again by a broad margin a little darker in colour.

The caterpillar is green, with white longitudinal lines, and its tail is forked.

The chrysalis, which is suspended by the tail, is of an angular shape, and has two sharp points on the head.

The markings of this species are very variable.

The figures are from specimens in my own collection.

HEATH BUTTERFLY.

PLATE XX.

Hipparchia Davus,	OCHSENHEIMER. CURTIS. STEPHENS.
Hipparchia Polydama, (var:)	STEPHENS.
Papilio Davus,	FABRICIUS. HAWORTH. JERMYN.
" *Tullia,*	HUBNER.
" *Philoxenus,*	ESPER.
" *Musarion,*	BORKHAUSEN.
" *Polydama,*	HAWORTH.
" *Polymeda,*	JERMYN.
" *Hero,*	DONOVAN. LEWIN.
" *Iphis,*	BORKHAUSEN. JERMYN. STEPHENS.
" *Tiphon,*	ESPER. (?)
" *Typhon,*	HAWORTH.
Maniola Tiphon,	SCHRANK.
Cœnonympha Davus,	WESTWOOD.
Satyrus Davus,	GODART.

I HAVE taken this insect in plenty on Thorne Moor, Yorkshire, in company with my good friend, J. C. Dale, Esq. Alas! it is some years ago; but "auld lang syne" has its pleasures too, and it would be wrong even to wish the "days that are gone" changed for those which may be yet to come. "What," said the M.P., "What has posterity done for us?" The "Pleasures of Memory" outweigh in the balance even those of "Hope."

The following localities are also given for it:—Cottingham, near Beverley, Yorkshire; the neighbourhood of Cromer, Norfolk; Delamere Forest, Cheshire; and between Stockport and Ashton, Risley Moor, Woolston Moss, near Warrington, White Moss, Chat Moss, and Trafford Moss, Lancashire; near the Lake of Derwentwater, in Westmoreland; Prestwich Carr, near Glandford Brigg, Lincolnshire; also in Northumberland and Cumberland; and on Barnet Heath. Also on Coates and Drumsheugh, Shehallion, Ben Nevis, Ben Lomond, Craig Chalaach, Ben More, the Isle of Arran, and the neighbourhood of Oban, in

Scotland, as also in the Shetland Islands; near the lakes of Killarney, and on the mountains of Donegal, in Ireland; Kinnordy, between Bala and Festiniog, in Wales.

The perfect insect occurs from the first week in June to the second or third in August.

The date of the appearance of the caterpillar is in May.

It feeds on the *Rhyncospora alba*.

The expanse of the wings in this species varies from an inch and a half to an inch and three quarters. The fore wings are of a fulvous brown colour, the fringe of a pale grey; near the outer corner are one or more eyes, following the line of the margin, all but one being in some specimens scarcely visible. The hind wings are rather darker than the fore ones, and there is a row of faint eyes following also their outer margin, at a little distance within it. An irregular band of a paler colour than the general tone of the wings runs more or less distinctly across them about the middle, following the same course as the eyes.

Underneath, the fore wings are nearly of the same colour as their upper surface. Near the outer corner are one or two dark eyes—a white dot in the centre, surrounded by a ring of black, and this by one of very pale buff. These eyes are followed by two or three smaller and less distinct ones, pale buff with a black dot in the centre. Within these is an irregular bar of still paler buff, wider at the upper than the lower part: it goes nearly, but not quite, across the wings.

The hind wings are greyish brown at the base, and as far as the middle, edged by an irregular very pale buff band, which runs nearly across the wing, the remainder of which is pale reddish brown, in the middle of which is an irregularly-waved row of eyes, each formed by a white speck surrounded by a black ring, and this by a pale buff one.

This is an exceedingly variable insect, and as may be supposed, several so-called species have been made out of one; permanent varieties seeming, as in the case of some other species, to belong to particular localities.

One described under the name of 'Polydama,' measures about an inch and a half in the extent of its wings; the fore wings are of a yellow brown colour, with two obscure eyes. The hind wings are brown, but with the inner edge broadly marked with dull white or pale buff, and there is a small obscure eyelet near the hind angle.

Another, described by the specific name of 'Typhon,' of the same size as the previous named ones, is described as follows:—On the upper side the wings are of a rusty grey or ochre-colour, brown at the base. The hind wings are generally darker, and without any distinct eyes. Underneath, the fore wings are dusky at the base, followed by an irregular whitish stripe, and the outer part greenish ash-colour, with from two to five small eyes, occasionally obsolete. The hind wings are greenish brown at the base, with an irregular interrupted bar, (this interruption forming the ground of the formerly supposed specific difference,) beyond the middle of the wing, and generally with about six small eyes, but their number is very variable; this bar is followed by a shade of greenish brown.

The female has the wings paler, and more tinged with ochre, with a large pale blot on each. The bar is succeeded by an ochre shade.

Mr. Westwood remarks on these different varieties, or supposed species, that in 'Davus' all the markings are complete, distinct, and unclouded; in 'Polydama' they are somewhat paler and less defined; and in 'Typhon' the broad band is divided into two irregular marks, while in further varieties some of the marks disappear altogether, and all are fainter. Also that 'Davus' has the little rings always more or less defined on the upper side, and is of a dull brown colour with a slight inclination to grey, the darker parts inclining to olive green. Typhon and Polydama have the little rings very slight, and in some instances altogether wanting on the upper side, while also the ground-colour is somewhat paler, and inclining to tawny, and on the under side all the markings are paler and less distinct.

The figures are from specimens in my own collection.

LEAST MEADOW BROWN.

SMALL HEATH.

PLATE XXI.

Hipparchia pamphilus,	OCHSENHEIMER. LEACH.
" "	STEPHENS. CURTIS. DUNCAN.
Papilio pamphilus,	LINNÆUS. LEWIN. DE GEER.
" "	HAWORTH. STEWART. HARRIS.
" *Nephele,*	HUBNER.
Cænonympha pamphilus.	WESTWOOD.

THIS is one of our commonest species, being abundant in almost all parts of the country in the summer time, "when the face of all nature looks pleasant and gay." It is frequent on heaths, as also in meadows and various other situations: it is, however, scarce in the neighbourhood of Falmouth, as W. P. Cocks, Esq. has informed me. Common at Brighton, Anstey, and Bisterne.

There are two broods, whereof the first appears the end of May and beginning of June, and the second in August and September, the last continuing till the middle of October.

The caterpillar is found in the beginning of May and August.

It feeds on the *Cynosurus cristatus*, or crested dog's-tail grass.

In this insect the wings expand to the width of from a little more than an inch to nearly an inch and a half. The fore wings are of a pale fulvous or tawny yellow colour, and the margins brownish, these being darker and more decided in the male than in the female. There is an indistinct eye near the tip, sometimes accompanied by a still smaller one, or by one or more black dots. The hind wings are of the same colour as the upper ones, but with a grey mark irregularly margined over their inner half, and with sometimes an obsolete eye near their lower corner.

Underneath, the fore wings are fulvous; their base, tip, and border, grey, and near the tip is a distinct eye—black with a white pupil,

and surrounded by a pale yellowish border or ring, in an indistinct patch of which the eye is placed. The hind wings underneath are greyish brown, with a faint tinge of reddish at the base, irregularly marked at the edge of the mark, as above; this is followed by a yellowish white streak of very sinuous shape, widest in the middle and narrower at either end. The remainder of the outside of the wings is the same colour as the base, but paler, and in it is a row of several minute faint eyes.

Varieties occur in which the eyes are more or less obliterated: I have one which has the corner of one of the fore wings curtailed in a curious manner, though possessing the rounded appearance.

ARRAN ARGUS.

ARRAN BROWN.

PLATE XXII.

Hipparchia Ligea,	OCHSENHEIMER. STEPHENS.
" "	DUNCAN.
Papilio Ligea,	LINNÆUS. SOWERBY.
" *Alexis,*	ESPER.
" *Philomela,*	HUBNER. ESPER.
Erebia Ligea,	DALMAN. BOISDUVAL.
Epigea Ligea,	HUBNER.
Oreina Ligea,	WESTWOOD.

THIS is a very rare butterfly, and has hitherto only been taken in the Isle of Arran, on the coast of Scotland. (Since the above was written, in the Isle of Mull also, I am informed.)

Sir Patrick Walker, and Alexander Macleay, Esq., captured specimens there. Those taken by the former gentleman, two or three in number, were met with near Brodrick Castle, and shewn by him to J. C. Dale, Esq. Many more will, I hope, yet be taken. " Floreat Entomolgia."

The expanse of the wings in this species is from a little under to about two inches. They are all of a rich dark brown colour, and the fore wings have a broad oblong patch of red near the outer margin, within which are four black eyes, the two nearest the tip being confluent. Their fringe is alternately brown and white. The hind wings have the like broad oblong patch of red, and in it three black eyes. Their fringe also is alternately brown and white.

Underneath, the brown of the wings is of a paler colour, and in the fore wings the band is of a brighter hue. In the hind wings it is almost obsolete, but beyond the middle is an irregular white band, between which and the hind margin are three black eyes with white centres, each surrounded by a red ring.

In the female, the eyes above have white pupils.

The caterpillar is described as being green, with a dusky line down the back, and some white lines along the sides.

SCOTCH ARGUS.

PLATE XXIII.

Hipparchia Blandina,	OCHSENHEIMER. STEPHENS.
" "	CURTIS. DUNCAN.
Papilio Blandina,	FABRICIUS. SOWERBY.
" "	DONOVAN.
Epigea Philomela,	HUBNER.
Oreina Blandina,	WESTWOOD.

THE British Butterflies are to be sought in various localities—the "Highways and Byways" of the country. Few of the latter will be found to exceed in the graces of quiet retirement, those of the district of Yorkshire, presently mentioned as one of the "habitats" of the present species.

This insect, formerly esteemed so rare, occurs in great profusion in the neighbourhood of Jardine Hall, Dumfriesshire, as I have been informed by Mrs. Hugh E. Strickland, and her sister, Miss Jardine. It is extensively distributed likewise in other parts of that county; also in the Isle of Arran; near Minto, in Roxburghshire, and about Edinburgh, and doubtless in most of the southern counties of Scotland. In Yorkshire a few have been captured at the foot of Whernside, in Craven, and Mr. Allis tells me that it is to be found in plenty near Grassington, in Wharfedale, also in Craven, on most of the hills and mountains of which district I fully expect that it will be discovered. It exists in profusion in one or two places not far from Newcastle, in Northumberland, and likewise in Castle Eden Dene, in the county of Durham, a sweet spot, well worth visiting for its own sake; as also at Grange, near Ulverstone, in Lancashire, in gardens.

The perfect insect appears about the beginning of August.

The Scotch Argus varies in the expanse of its wings from an inch and a half to two inches. The upper side is of a uniform dark bronzed brown colour, the fore wings having a dark orange-red patch near the tip, wider in its fore than its hind part, which two are sometimes divided by a constriction in the middle of the patch. In the upper part of this patch are two black eyes, each with a white speck in its centre; in the hinder part of the patch is one similar eye, but smaller in size, and occasionally it is obliterated. In some specimens there are as many as five eyes.

The hind wings, also of the same dark bronzed brown, have a waved or indented bar, or united series of round marks of dark orange-red following the outer margin, a little distance within it, and in it are generally three small black eyes with white pupils, and a black dot in their outer part. In some specimens there are only two eyes. The fringe of all the wings is brownish, but darker in the male than in the female.

Underneath, the fore wings are of much the same general colour as on the upper side, the brown bar shewing through as above, but of a more yellow tinge, and the eyes in it similarly appearing. The hind wings have a tint of grey with the brown at the base, which is succeeded by a broad waved brown bar of the general colour of the wings, this, by another grey wave of the colour of the base, in which are sometimes a few rudimentary eyes, but in other specimens it is quite plain, and the margin of the wing again is brown. The colour of these bars varies very considerably in the males and females, and also according to the locality in which the insect is found.

The caterpillar is described as being light green, with brown and white longitudinal stripes, and the head reddish.

It feeds on various kinds of grass.

The eggs are said to be of a whitish colour, specked with brown.

This species varies much.

The figures are taken from specimens in my own cabinet.

SMALL RINGLET.

MOUNTAIN RINGLET.

PLATE XXIV.

Hipparchia Cassiope,	OCHSENHEIMER. STEPHENS.
" "	CURTIS. DUNCAN.
Melampias Cassiope,	HUBNER.
Papilio Æthiops minor,	VILLARS.
" *Melampus,*	ESPER.
" *Mnemon,*	HAWORTH.
Oreina Cassiope,	WESTWOOD.

THIS is a rare species in this country, or even if eventually proving to be more plentiful than has hitherto been supposed, a very local one, being only to be met with, so far as at present is known, on the mountains of Cumberland and Westmoreland, from whence you look down on the scenery, which, in even a more than ordinary manner, gives ground for the exclamation that "God made the country, and man made the town."

The following are localities for it:—Rannoch, in Perthshire, Helvellyn, Red Skrees, a mountain near Ambleside, and Gable Hill, and Stye Head, between Wastwater and Borrowdale.

The earliest date of its appearance is about the 11th. of June, but the latter end of that month is the proper time. The females are later out than the males, and have been taken in July and August.

The caterpillar feeds on the mat-grass, (*Nardus stricta.*)

The wings expand to the width of one inch and a quarter, or a little over. They are of a very dark brown colour, and the fore wings have a red bar near the outside, interrupted by the nerves. In this bar are generally four small black dots with obscure pupils, but some specimens have only three, or even only two eyes, while some are without them altogether, and in others the bar itself is reduced to a few red spots.

The hind wings have also a red bar near the outside, with three eyelets like those on the fore wings.

Underneath, the fore wings are reddish brown, with the red band, which, as it were, shews through, marked with four black spots. The hind wings are grey or brown, with a metallic tint, and have three black spots, each surrounded by a narrow red ring.

SILVER-BORDERED RINGLET.

PLATE XXV.

Hipparchia Hero.	OCHSENHEIMER. STEPHENS. CURTIS.
Papilio Hero,	LINNÆUS. HAWORTH.
" *Sabæus,*	FABRICIUS.
" *Melibæus,*	ERNST.
Cœnonympha Hero,	WESTWOOD.

THIS is one of the rarest British insects; two specimens only being all that at present are known so have occurred. One, a female, was taken by Mr. Plastead, near Withyam, on the borders of Ashdown Forest, Sussex, and Mr. Stephens states that he obtained another from the neighbourhood of Lamberhurst, in the same county. Who knows however, how many may have been overlooked, or how many may yet be taken by keen observers, who act upon the moral conveyed in Mrs. Barbauld's instructive story of "Eyes and no Eyes!"

This butterfly measures about one inch and a half in the expanse of its wings; they are of a general fulvous brown colour. The fore wings are paler along the front edge, and have an orange stripe close to the hind margin, near which are two small indistinct orange-coloured eyes with brown centres.

The hind wings have also a narrow orange stripe near the outer margin, above which are four large black eyes, with minute whitish-coloured pupils, and surrounded by a broad orange ring.

Underneath, the fore wings are of the same general colour as the upper side, but there is a narrow silver stripe adjoining the orange one. In the hind wings there is a continuation, as it were, of the silver stripe through them, and also an irregular whitish bar rather beyond the middle, succeeded by orange, in which are seven eyes of different sizes, the two nearest the inner corner being the smallest, and confluent, the eye being black, with a white pupil, and surrounded by a rim of orange. I imagine that probably these eyes are variable both in number and size.

WHITE ADMIRAL.

PLATE XXVI.

Limenitis Camilla,	Leach. Curtis. Duncan.
" "	Hubner. Westwood.
Papilio Camilla,	Linnæus. Haworth.
" "	Lewin. Donovan. Harris.
" *Sibilla,*	Fabricius. Stuart.

This most elegant and graceful species is of decidedly local distribution, though in several places it occurs in tolerable plenty—the "Happy hunting grounds" of the entomologist. It is found in shady places in the retired depths of woods, where the sun gleams through at intervals. It is fond of alighting on the bramble blossom, the nectar of which it sips. R. P. Postans, Esq. informs me that it is taken in the greatest abundance in Hartley Wood, near St. Osyth, Essex; and also, though more sparingly, near Colchester and Dedham, in the same county. It is also found near Rye, Sussex; in Coombe Wood, near London; in a wood near Parley Heath, abundantly in woods near Winchester, and in the New Forest, Hampshire; Haslemere, Surrey; near Peterborough, and in Sywell Wood, near Northampton, and I believe at or near Lilford, in the county "of that ilk;" Ipswich, Suffolk; Enborne Copse, near Newbury, Berkshire; in Birch Wood, Kent; near Finchley, Middlesex; and one specimen of late years in the Isle of Wight, where M. A. Bromfield, Esq. says that it used to be common in the woods near Ryde. J. Wesley, Esq. took several and saw others there in 1855. It also occurs near Weston-super-Mare, in Somersetshire; likewise in Woolmer Forest.

This butterfly appears the second week in July.

The caterpillar feeds on the honeysuckle.

The expanse of the wings is from two inches and a quarter to two inches and a half; the upper surface of the fore wings is dull brownish black, with a curved band of interrupted large white spots, rather outside the centre, the middle one being very much smaller than the

others. There are two other small white spots near the corner, forming, as it were, the right boundary of the white bar, the rest of it being filled up with the general colour of the wing; between the bar and the base of the wings is another dull white mark. There are also one or two small faint white oblong spots near the middle of the outer margin of the fore wings, and they are fringed with white.

The hind wings have the same white bar continued through their centre, narrowing to the end: they are also fringed with white. There is an obscure red spot, within which are two black dots, near the inner lower corner. Inside the fringe is a band of a darker colour than the ground of the wings, and within this two others of interrupted spots of the same, and two others are obscurely visible within the white bar.

Underneath, the general ground-colour of the fore wings is fulvous red, and all the white marks from above shew through, and the dull ones are all clear. The centre of the wings about the lower part of the bar has a tinge of ash-colour and faint bronze. The fringe is alternate brown and white, the latter being intersected by the former being waved. This is succeeded by another curved line, leaving three small crescents of white, and a larger one again within the uppermost of the three, these two last being the ones on the upper side that shew through; after this the line fades nearly away in the fulvous ground-colour of the wings. There are four short dark waved across lines, two on each side of the white spot between the bar and the base.

The hind wings have two rows of blackish brown dots between the bar and the margin, then one or two indistinctly defined white marks, then a row of crescent-shaped white marks, very faintly discernible, towards the outer corner, and then the white margin, indented with the fulvous brown of the wings, shewing through between it and the last-named row of white crescents in the shape of a waved line.

The caterpillar is green, with the head and legs reddish.

The chrysalis is greenish or brownish, with golden spots. It has a large and prominent appendage on the back, and the head is divided into two forward projections.

Two varieties of this insect have been taken near Colchester, in each of which the white spots on the wings were nearly effaced, the white band entirely or nearly obliterated.

RED ADMIRAL.

ALDERMAN BUTTERFLY.

PLATE XXVII.

Vanessa Atalanta,	FABRICIUS. STEPHENS. CURTIS.
" "	DUNCAN. WESTWOOD.
Papilio Atalanta,	LINNÆUS. HAWORTH. LEWIN.
" "	DONOVAN. WILKES. ALBIN. HARRIS.
Ammiralis Atalanta,	RENNIE.
Pyrameis Atalanta,	HUBNER

THIS magnificent butterfly, one of the richest coloured of our native species, is met with throughout Europe, and also in the northern part of the continent of Africa. It is widely distributed in England, but, according to W. P. Cocks, Esq., is scarce in the neighbourhood of Falmouth. It occurs near Looe. I have seen it most abundantly in Worcestershire. It ranges from Brighton to Bisterne. In Yorkshire also plentifully near Nunburnholme and Nafferton. In Wales at Llandudno.

It is frequently to be met with in gardens, being fond of the flowers of the dahlia and the blossoms of the ivy, and is a very bold and fearless species, so as to be for the most part easily approached. A pleasant sight it is to watch it in your quiet retirement in the country, where, "the world forgetting, by the world forgot," you can enjoy in tranquillity the "Thousand and one" beautiful sights in which the Benign Creator displays such infinite wisdom of Almighty skill.

The perfect insect appears in June, July, and August, and many individuals live on to the winter, and even survive until the following spring, when they again appear, and, though faded from their former

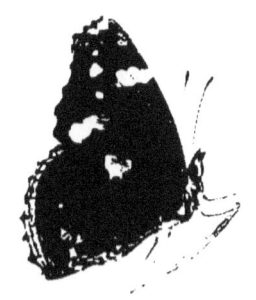

splendid beauty, still glitter as welcomely in the eyes of the entomologist as the sunshine of the returning spring that calls them out. One which had lived through the winter was seen on the 2nd., 10th., 12th., and 18th. of June.

The caterpillar is to be found in July. It continues fourteen days in chrysalis.

It feeds on the common nettle, giving a preference to the seeds.

The wings of this insect expand to the width of from two and a half to three inches. The ground-colour of the fore wings is intense velvet blue black on the outside half, and intense copper brown black on the inner. A bar of lovely red runs nearly across them, not quite reaching to their lower corner; towards which it is slightly curved. It is irregular on its margins, formed, as it were, of a series of rounded oblong patches. Between this bar and the tip of the wings is a similar short bar of pure white, formed of three patches; and a fourth lies like an island beyond it, interrupted by the ground-colour of the wings. This latter one forms the largest of a chain of white spots, which sweep by it, one inside, and three outside, of which the latter two are very small dots, and the third at the top is a narrow oblong curve. Beyond these, and between them and the tip of the wing, is an obscure wave of purple blue. The margin of the wings is white, indented crescent-wise on the ground-colour.

The hind wings are deep velvet brown black, with a broad margin to all their middle part of fine red, in which are four black dots. At the lower inside corner of the wings are two small conjoined oblong marks of purple blue. The margin of the wings is white, indented in curves into the red bar, and shaded with blackish brown, forming a sort of dots between each segment.

Underneath, the fore wings are mottled brown at the tip, and there are near it two small white dots, surrounded by two rings of brown. The red bar and the white spots shew through from above, and between them, near the fore edge, are two small waved transverse stripes of metallic blue. The upper inner corner of the red bar is intersected by a bar of brownish black, and bounded by another, within which again is another cross mark of metallic blue. The fringe of the wings is white, indented as on the upper side. The hind wings are most beautifully marbled, mottled, and variegated all over, in a manner that hardly admits of description, with black, brown, buff, bronze, and grey, with a rather large triangular yellowish white mark, with some brown in its centre, in the middle of their front margin, and two heart-shaped marks some way within their hind margin, somewhat like the eye of a peacock's feather, brown, margined with black, and with a metallic green eye. These wings are also edged with white in the same crescent-shaped way.

The caterpillar is of a dusky green colour, with a yellowish line along the back, and a pale line on each side above the feet.

The chrysalis is blackish or brownish above, and underneath, grey, with golden spots.

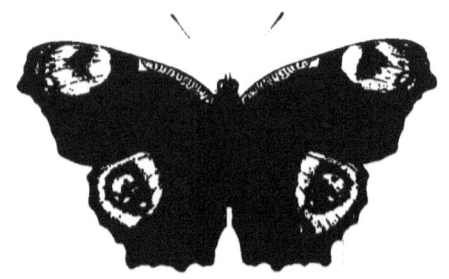

PEACOCK.

PLATE XXVIII.

Vanessa Io,	FABRICIUS. OCHSENHEIMER.
" "	STEPHENS. DUNCAN. WESTWOOD.
Papilio Io,	LINNÆUS. HAWORTH. LEWIN.
" "	DONOVAN. ALBIN. WILKES. HARRIS.
Inachis Io,	HUBNER.

As the student in entomology, or indeed in any branch of Natural History, meets for the first time with one new species after another whose distinctive appearance it had never even come into his mind before to conceive, he repeatedly exclaims, not indeed perhaps in words, but in the admiration of his mind, "Wonders never cease:" well do I remember the intense pleasure which, when a boy, the first sight of the Peacock, the Red Admiral, and the Brimstone afforded me. I wish others to experience the same gratification, and shall be truly happy if my "History of British Butterflies" furthers the cause of the gladsome science which it is intended to illustrate.

This splendid species is common throughout the greater part of the country, though less so as you advance farther north—from Brighton to Broomsgrove, in Worcestershire, and Anstey, in Warwickshire. In the south of Scotland it is but sparingly met with.

The perfect insect appears in the middle of July, and by no means unfrequently survives until the following spring, hybernating during the winter in sheltered "nooks and corners."

The caterpillar is found in the beginning of July.

It feeds on the common nettle.

In this grand fly the wings expand to the width of from two and a half to three inches; the fore wings are of a rich dark brownish red; on their front margin there are two black nearly triangular-shaped marks, the inner one smaller than the other, which latter forms the inside of a large patch, angular on its inner side, and rounded on its outer one, which is, as it were, partially eclipsed by a large eye, whose ground-colour is yellowish buff, and within whose orbit are marks of black and purple red, with a border of blue spots, and three pale blue specks, followed by two others outside it. It is much in the form of the handle of "Charles's Wain," the always well-known constellation that guides the traveller by shewing him unerringly the north, a beacon which, as the church spire points upwards to raise the mind towards Heaven for the journey thither, directs by a downward indication the earthly pilgrimage of many a benighted wanderer both by sea and land. "God is great," says the Moslem, and verily the Musselman speaks true in his saying. Great He is, in the stars of Heaven, those unknown worlds of illimitable space, and great, equally great, in the beautiful though humble insect before us.

The outer margin of the wings is brown, and their front edge is striated on the inner half with streaks of dark yellowish and black.

The hind wings are also reddish brown on the central and hinder parts of their surface; the base being brown, studded with innumerable specks of yellow dust, and the outside border brown: near the outer corner is a very large eye surrounded with a colour which approaches more nearly to white than any other, and it is bounded on its inside with blackish brown. The eye itself is black, with five blue specks in it—two, two, and one.

Underneath, the fore wings are dark brown, streaked across with an infinity of darker marks, some wider, some narrower, and some of deeper shades than others.

The hind wings are of a darker ground-colour than the fore ones, striated in the same way, and across their centre is one large waved bar formed by dark edges, and in its centre an obscure yellowish white dot.

The caterpillar is gregarious in its habits, black, spined, spotted with white, and the hind legs are red.

The chrysalis is indented, of a greenish colour, and dotted with gold. It remains exactly a fortnight before the butterfly "comes out."

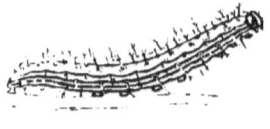 

LARGE TORTOISE-SHELL.

PLATE XXIX.

Vanessa polychloros,	OCHSENHEIMER. CURTIS.
" "	STEPHENS. DUNCAN. WESTWOOD.
Papilio polychloros,	LINNÆUS. HAWORTH. LEWIN.
" "	DONOVAN. ALBIN. WILKES.
Eugonia polychloros,	HUBNER.

THIS is a very fine species, though painted with no particularly gay colours: it is at the same time sufficiently common. It frequents woods, lanes, and gardens.

I have seen this insect in the parishes of Bossall and Huttons Ambo, Yorkshire, and have taken it in a wood a few miles from Worcester, in which county it is tolerably plentiful. I also once procured the larva, and reared it to the perfect insect, at Charmouth, Dorsetshire. It is common in the neighbourhood of Feversham and Milstead, near Sittingbourne, Kent, where the Rev. Henry Hilton has taken it; and occurs at Sywell Wood, near Northampton, Barnwell and Ashton Wold, and the neighbourhood of Polebrook and Lilford, Northamptonshire. In the woods of Suffolk and Essex it is very plentiful, at least in those near Stoke Nayland and Colchester. It occurs also near Great Bedwyn and Sarum, Wiltshire; plentifully at Bradfield, Berkshire, where my second son, Reginald Frank Morris, has taken it in the caterpillar state; in the neighbourhood of London; and also at Ely, and other parts of Cambridgeshire; in Norfolk occasionally, as Mr. Robert Marris informs me; and, though rarely, in the woods on the banks of the River Dart, in Devonshire. In Hampshire it is tolerably plentiful near Winchester, J. Wesley, Esq. informs me; also at Bisterne; so to at Brighton, Sussex; and Anstey, Warwickshire.

In Scotland, it has been found as far north as Dunkeld, and in other places to the south of it.

This butterfly comes out in the middle of July, but some individuals, hidden away probably in some sheltered corner, survive through the ungenial season, " when the stormy winds do blow," and re-appear for a time the following spring. It is rather uncertain in its appearance, being much more plentiful some years than others; but indeed with what insect is this not the case?

The caterpillar feeds on the elm, and at first the whole brood are gregarious, being associated together until their first change of skin, under a common web.

The wings measure from two inches and a quarter to three inches in expanse. The fore wings are fine rich orange brown, dusky at the base. They have four black patches of different sizes on the front edge, the farthest forming the commencement of a streak, which follows the windings of the margin of the side at a little distance within it. The margin itself is also dark. The next patch is the head also of a line of dots, four or five in number, across the wing, running inwards, the lowest being a large one, and between it and the streak already spoken of is another dot. The hind wings are of the same fulvous colour as the upper ones, but their inner portion is more extensively dusky. There is a large triangular-shaped black patch at the centre of their upper edge, forming indeed the boundary of the dusky part. Their outer edge is indented and paler than the rest, and within it is a furbelow of dark blue crescents, ending above in blackish, margined with a blackish blue one, the middle ones being the largest, and the horns of the crescents running through the margin to the extremity of the wing.

Underneath, the fore wings are dull brown, dark at the base, then lighter, and then darker again, edged interiorly with very dark blue at the central and lower part of the outside, the light part thus assuming the form of a band. The hind wings are marked precisely in the same way, the dark and light parts being continuations of those on the fore wings. There is a small white dot near the centre, in the outer part of the dark base.

The caterpillar is blackish or brownish, with a yellow line along the side, and yellow spines.

The chrysalis is rather dark greenish brown, with small golden spots. It is frequently found beneath the coping of walls, underneath the trees on which the larva has fed, as well as attached to the tree itself.

There are several varieties of this species, the black markings being more or less diffused.

The figure is taken from one in my own cabinet.

SMALL TORTOISE-SHELL.

PLATE XXX.

Vanessa urticæ.	FABRICIUS. OCHSENHEIMER. STEPHENS.
" "	DUNCAN. WESTWOOD.
Papilio urticæ,	LINNÆUS. LEWIN. DONOVAN.
" "	ALBIN. WILKES.

THIS is one of our most common species, and therefore but little thought of in comparison with others of greater rarity. It is, however, a handsome insect, and in its general markings very much resembles the Large Tortoise-shell, though the difference of colour instantaneously distinguishes the two.

It is plentiful at Brighton, Charmouth, Anstey, Bisterne, Herne Bay, and Broomsgrove, etc.

The perfect insect, there being two broods, appears in the end of June and beginning of July, and the latter end of August or September. The second brood often survives the rigours of even our northern winter, and is seen again the following spring, flitting gaily among the early flowers of the garden, or along the grassy "Banks and Braes," and anon borne away by some fitful breeze of the uncertain season.

The caterpillar is to be found in the beginning of June, and again in the middle of August. It is gregarious in its habits, principally in the earlier stages of its growth.

It feeds on the nettle.

This butterfly varies in the expanse of its wings from an inch and three quarters to two inches and a quarter; the fore wings are of a rich red orange colour, but the base is dark. There are three large black patches on their front edge, and between these the ground-colour is much paler than on the general surface, being light yellowish orange; beyond the outermost one is a white triangular-shaped mark. Near the base of the middle part is a large irregular spot, and above this in the direction of the outer corner, two smaller ones; the outer edge is

K

dark buff, followed immediately by a black indented stripe, in which are a series of small dark blue crescents.

The hind wings are also rich orange red, but all the base is dark coloured, and they are bordered with dark buff, followed by an indented black band, in which is a row of dark blue crescents of larger size than those in the fore wings, leaving the orange as a bar across. Underneath, the markings are the same, but the orange is changed to stone-colour; the margins are the same, but darker, and separated from the rest by an indented line of metallic blackish green.

The lower wings have the bar replaced by a darker stone-colour; the margins separated by a row of crescent-shaped dark blackish green spots.

The caterpillar is of a dull colour—a mixture of green and brown, with paler lines down the back and sides, and beset with black spines: the head is black.

The chrysalis is brownish, with golden spots on the fore part, and sometimes nearly entirely golden.

In varieties of this species the black spots have been more or less enlarged or diminished, so as in some cases to be confluent, and in others obsolete. In one figured by the Rev. W. T. Bree, of Allesley, in the "Magazine of Natural History," the second and third black bars on the front edge are united, and the two round spots on the same wings are absent, the hind wings being uniformly obscure.

A very singular 'Lusus naturæ,' preserved in the cabinet of Mr. Stephens, has occurred in the Small Tortoise-shell, Mr. Doubleday having taken one near Epping, with five wings, the fifth, of small size, being affixed to one of the hinder ones, whose markings it repeated.

(A hymenopterous insect with seven legs, four on one side and three on the other, and still preserved in the cabinet of J. C. Dale, Esq., was captured several years ago by my brother, Frederick Philipse Morris, Esq., in a wood near Axminster, Devonshire.)

The engravings are from specimens in my own collection.

CAMBERWELL BEAUTY.

WHITE BORDER. GRAND SURPRISE.

PLATE XXXI.

Vanessa Antiopa,	OCHSENHEIMER. STEPHENS.
" "	DUNCAN CURTIS.
Papilio Antiopa,	LINNÆUS. HAWORTH. LEWIN.
" "	DONOVAN. HARRIS. WILKES.
Eugonia Antiopa,	HUBNER.

THE wide uncertainty of the periodical appearance of this very fine butterfly in our country is extremely remarkable, and "whither away?" between the dates of its visits is a question we cannot answer. About eighty years ago it appeared in immense numbers. Again, in the year 1819, it was observed in abundance in all parts of the kingdom: comparatively few have been seen since, but latterly, within the last few years, more have been met with, probably from having been better looked after.

The neighbourhood of Rawmarsh, near Rotherham, Yorkshire, is one of the most uniform localities for this rare insect that I am aware of. It has also occurred near Huddersfield; so, too, at Lockington, near Beverley, from whence I received a fine specimen alive by the post, about the year 1856, the donor being an entire stranger to me, and requesting me to acknowledge its due receipt to "an old woman, post-office, Lockington," which I accordingly did, with many thanks. Also at York, Whitby, Allerthorpe Common, and Langwith. In my own garden at Nafferton Vicarage, near Driffield, one, as there seemed no doubt that it must have been, was seen by my servant a few years since. It has occurred also near Cromer, Norfolk, and one was taken near Yarmouth, in August, 1852, as Mr. M. C. Cooke has told me. I thought that I once saw one, three or four miles south-west of the city of Worcester. It has also been captured in the following localities, the year 1846 having been unusually productive of the species:—In that year, in a garden in the suburbs of York, one was taken, (as was another in 1852, four in 1858, and yet another in 1859,) and two others were seen at the same time; one near Epping, on the 12th. of September, and another was seen; one taken near Yaxley about the same time. One at Winchester, Hampshire, near some willows, on the 14th. of September; two on a mulberry tree in the garden of the vicarage at Stowmarket, Suffolk, and another in the same neighbourhood about the same date; one near Ipswich by Mr. Charles Eaton, on the 30th. of August, 1852, and another on the 31st. of the previous year; one in a garden at Lincoln, in August; one at Herne Hill, Camberwell, in a garden on the 12th. of September; one at Kensington, on some ripe fruit, on the 21st. of September; one at Tottenham, in September; and one seen at Streatham, resting on the sill of a window; one at Weymouth, in Dorset. Two were seen at Clapham, Bedfordshire, one on the 13th. of August, and the other a few weeks later on an apple tree; another near Woburn, in the same county; and one was caught at Bronstone Wilderness, near Leicester.

One was seen near Ventnor, in the Isle of Wight, on some fallen peaches, the 23rd. of August, and two following days; another near Sea View, about two miles from Ryde, about the same time; and another between Kingston and Leatherhead, Surrey, probably the same that was shortly afterwards taken at Mickleham. One was taken at the West India Docks, London, on the 3rd. of September, and another at Limehouse about the same time; one near Mickleham; one at Southwell, Nottinghamshire; one at Saffron Walden, Essex, early in September; and some taken and others seen in different parts of Norfolk, resorting to the blossoms of the ivy. One was taken on the 5th. of September, 1846, in the garden of Widmore House, Bromley, Kent, by A. Henry Taylor, Esq., who, I am informed, saw several others also on the same occasion; and one at Seaton Mill, near Workington, Cumberland, shortly after leaving the chrysalis, on the 10th. of September, 1852, of which Mr. Thomas Jackson, of that place, has informed me. In previous years one was taken near Nottingham; one near Stoke-by-Nayland, Suffolk; two near Colchester, and one between Dedham and Colchester, Essex, in the month of August; one at Cromer, in Norfolk, in the year 1817, by H. Barclay, Esq.; and formerly, whence its name, at Camberwell.

In Scotland it has been noticed near Coldstream, Berwickshire, and so far north as Ayrshire.

The butterfly appears in the beginning of August, and, like others of its class, occasionally survives through the winter, and re-appears, after its long sleep, with the advance of the new year.

The caterpillar feeds on the willow and birch, and is said to be found on the topmost branches.

This butterfly varies in the expanse of its wings from a little under three inches to three inches and a half. The fore wings are of a fine dark rich claret-colour, margined with dull white, or yellowish. Inside the margin is a row of blue spots, on a velvet black ground. The hind wings are of the same dark claret ground-colour.

Underneath, the wings are ash brown, with a great many slender transverse black lines; the white margin and spots shew through, as do the bar and the blue spots, but only faintly, if at all.

The caterpillar is gregarious, black in colour, with spots on the back, and some of the legs of a red colour.

The chrysalis is dull black, with fulvous spots, and dentated in appearance.

The engraving is from a specimen in my own collection.

COMMA.

PLATE XXXII.

Vanessa C-album,	OCHSENHEIMER. CURTIS.
" "	STEPHENS. DUNCAN. WESTWOOD.
Papilio C-album,	LINNÆUS. LEWIN. DONOVAN.
" "	HARRIS. ALBIN.
Polygonia C-album,	HUBNER.
Comma C-album,	RENNIE.

THIS very handsome and singularly-shaped species has been noticed very abundantly two successive years by Mr. Graham, of York, near Green Hammerton, Yorkshire, alighting in hundreds on the blossoms of the common wild Scabious; and James Dalton, Esq. has taken it at Hackfall, so celebrated for its beautiful scenery. It also occurs near York. I have seen it occasionally near Doncaster and Burnby, also in the garden of Nunburnholme Rectory, and by the side of the Brant Wood, also at Howsham, Sutton-on-Derwent, Langwith, and Buttercrambe Wood, and in other parts of the same county. I have also taken it in plenty in the neighbourhood of Broomsgrove, Worcestershire, in all parts of which county it is doubtless to be found, as it also is throughout Warwickshire. Other localities for it are Sywell Wood, near Northampton, Lilford, Barnwell, and Ashton Wold, and the neighbourhood of Polebrook, Northants. It used formerly to be taken at Glanville's Wootton, Dorsetshire, by J. C. Dale, Esq., but is now never seen there. It is very rare near Stoke-by-Nayland, Suffolk, and very abundant near Bristol, and doubtless in other parts of Gloucestershire. It occurs at Brighton, Lewes, and Anstey, also on Ranmore Common, near Dorking, Surrey. In Wales, near Llandudno. In Hertfordshire it also occurs, and the neighbourhood of London; and in Scotland, in Fifeshire.

The perfect insect appears in June, and in August or September,

there being two broods in the year, the latter of which is paler in colour than the former. Some individuals live throughout the winter, and re-appear the following spring in the sunshine, too soon to be obscured by the clouds which the "Wandering Winds" of that early season unwelcomely interpose.

The caterpillar feeds on the elm, willow, honeysuckle, and other trees, and the hop, nettle, and other plants.

The wings expand to the width of from an inch and three quarters to rather over two inches. The fore wings are a beautiful rich fulvous orange colour; the outer margins are dark orange brown, darkest in the middle, and lighter at each extremity. There are three black patches on the front edge, the outer one fading into the fulvous brown of the border, the next lighter brown at its upper edge, and the next some little way within the extreme margin; beneath these are one larger and two smaller black spots of irregular shapes. The hind wings are dusky at the base and at the outer corner, their ground-colour is also fulvous orange, the border darker, edged with cream-colour. Inside it is a row of pale buff crescents, forming as it were, the centre of a band of a darker brown than the rest of the wing, which, divided by them, leaves a series of blots following the shape of the margin. On the centre of the fore edge is a large black patch, and beneath it, on the inner side, another smaller one.

Underneath, the fore wings are elegantly variegated with transverse marks of rich brown, grey, whitish grey, and metallic green, in which latter are small black specks. The hind wings are marked in much the same way, with a white C in the middle, whence the name of the insect, both scientific and vernacular.

The whole under side varies very much in different individuals, and in the spring and autumnal broods. In some it is almost wholly of a uniform dull metallic bronze brown colour. In others, the border is of an exceedingly rich brown, and the whole surface much variegated.

The caterpillar is of a brownish red colour; the back reddish in front, and the hinder part white; it is remarkable for the sides of the head having two projections, which are bristled, as are also the spines on the body.

The chrysalis is pale brownish red, and spotted with gold. It remains in this state about fourteen days.

Mr. Westwood observes how this species is subject to an extraordinary variation in the form of its wings, the incision in the outer margin of the fore wings being in some specimens so deep that it forms nearly a semicircle, whilst in others it is scarcely more than a sextant, the other indentations being equally varied.

I once had a singular variety taken near Doncaster, and which I gave to J. C. Dale, Esq., in which all the black spots on the fore wings were run into one large patch.

The plate is from specimens in my own collection.

ALBIN'S HAMPSTEAD EYE.

PLATE XXXIII.

Cynthia Hampstediensis,	STEPHENS. WESTWOOD.
Hipparchia Hampstediensis,	JERMYN.
Papilio oculatus Hampstediensis ex aureo fuscus,	PETIVER.

THE only specimen of this insect that has ever yet been recorded, was captured at Hampstead, near London, by Albin, and then first described and figured by Petiver. It has since been continuously figured and described by succeeding entomologists, who have faithfully copied the original picture. By some it has been considered a foreign specimen, accidentally imported; by others as the product of two different species. The specimen is however no longer in existence, and cannot speak for itself; no 'Ecce signum' can now testify to the truthfulness of the entomologist who shall pretend more accurately to describe it, than in the stereotyped form which has come down to the present day.

The fore wings have been described as fulvous brown, with three transverse dark brown markings; two lengthened ones near the hinder margin, and the margin itself yellow: there is a large eye near the tip and another near the lower corner. The hind wings are also brown with a yellow margin, and with two large eyes following the margin.

Underneath, the fore wings are yellowish brown, with brown cloudings, and a row of brown crescents near the margin. The hind wings are dull yellowish brown, with darker cloudings of brown at the base, a small eye near the corner, and a row of four brown spots, between which and the margin is a nearly obsolete row of brown crescent-shaped marks.

PAINTED LADY.

PLATE XXXIV.

Cynthia cardui,	KIRBY. STEPHENS.
" "	DUNCAN. WESTWOOD.
Vanessa cardui,	GODART. LATREILLE.
" "	MEYER. HUBNER.
Libythea cardui,	LAMARCK.
Papilio cardui,	LINNÆUS. FABRICIUS. HAWORTH.
" "	LEWIN. DONOVAN. SHAW.
" "	PANZER. WILKES. ALBIN. HARRIS.

THIS is, I believe, one of the most universally-distributed species of butterfly in the world, being found in every quarter of the globe, and in every, or almost every country, both the hottest and the coldest; in Europe and North and South America, New South Wales and Java, North and South Africa. It is, however, very uncertain in its appearance, at least in any numbers.

In Yorkshire I have taken it not unfrequently at Nafferton, also at Nunburnholme; near Falmouth in was plentiful in the year 1849, but scarce in 1850 and 1851: it has also been taken at Looe, and occurs in the Isle of Wight. In 1850 not one was noticed by R. B. Postans, Esq., near Stoke-by-Nayland, while in 1851 it was to be seen in extreme abundance there. In the same year it was captured near Hunstanton, Lynn, and other places in the county of Norfolk, and in Cambridgeshire, as Mr. Robert Marris has informed me. Also near Brighton. In Wales at Llandudno.

In Ireland it is abundant near Ardrahan, in the County of Galway; so A. G. More, Esq., of Trinity College, Cambridge, has written me word.

The perfect insect is found in June, July, and also at the end of August and beginning of September—a second brood. It occurs in some years in great plenty, and in others is but rarely seen. In the year 1828 an immense swarm passed over part of Switzerland in such

vast numbers, that their transit occupied several hours, in a manner 'mirabile dictu.'

The caterpillar is found in the middle of July.

It feed on the thistle, *(Carduus lanceolatus,)* the nettle, *(Urtica dioica,)* the mallow, *(Malva sylcestris,)* and the artichoke, *(Cynara scolymus.)*

The wings expand to the width of from two inches and a quarter to two and three quarters, or a little over. The fore wings are brown at the base, the tip blackish, with five white spots, the largest of which adjoins the front edge, and the four others—two small ones between two larger, the upper of which latter also adjoins the front edge—form a curved line, between which and the margin is a slender interrupted whitish line, the margin itself being white, indented crescent-wise. The rest of these wings is fulvous orange, with a suffusion of pink more or less extended from the upper inner part, and with three indented black united spots. The hind wings have the base and the inner margin brown, the remainder fulvous, with many black marks arranged in three rows, the inner one forming a series of round larger and smaller spots, and the outer one adjoining the margin, which is whitish.

Underneath, the fore wings are marked nearly as above, but the tawny colour is more diffused; the dark spots are smaller, and the tip of the wing is dark stone-colour, instead of black. The hind wings are mottled in the most charming manner with pale olive brown, yellowish buff, and white, the veins being white. Near the hind margin is a row of slender blackish blue marks, above which are four eyes, the two middle ones being smaller than the two outer ones, which are encircled with black.

The caterpillar is solitary, of a brown colour, with interrupted yellow lines along the side, and spined.

The chrysalis is brown, with ash-coloured lines and golden spots.

This butterfly varies considerably both in size and in the amount and depth of the pink colour on the wings.

SCARCE PAINTED LADY.

HUNTER'S CYNTHIA.

PLATE XXXV.

Cynthia Huntera,	KIRBY. WESTWOOD.
Vanessa Huntera,	DALE. STEPHENS.
Papilio Iole,	CRAMER.
" *cardui Virginiensis,*	DRURY.
" *Huntera,*	FABRICIUS. HERBERT.
" "	ABBOT AND SMITH.

ONE single specimen of this American species has, as yet, been obtained in this country. It was taken by the late Captain Blomer—no relation of "Mrs. Bloomer,"—at Withybush, near Haverfordwest, South Wales, in July or August, 1828. It was kept by him for some time as a variety of the Painted Lady, and subsequently presented to J. C. Dale, Esq., of Glanville's Wootton, Dorsetshire, in whose cabinet I have seen it.

This fly appears periodically in abundance in its native country, while in some seasons it is rare.

The caterpillar is said to be found at the end of April, or the beginning of May, and at the end of July, or beginning of August, so that, if so, there are no doubt two broods in the year.

It is variously said to feed on the wild balsam, and the obtuse-leaved cudweed, (*Gnaphalium obtusifolium.*)

This insect measures about two inches and three quarters in the expanse of its wings, which are of a tawny orange and brown colour. The fore wings are brown at the base, and have several irregular blackish bars, and the tip blackish, with a long white spot, and four dots near the tip also white, between which and the margin is a pale interrupted waved streak. The hind wings have a slender interrupted brown line near the edge, succeeded on the inside by

four more or less distinct eyelets, and two very slender dark lines near the margin, which is lilac-coloured.

Underneath, the fore wings are elegantly varied with white, orange, brown, and pink, with two eyes near the tip. The hind wings are brown near the outside, with two very large eyes margined with black; and between these and the margin are waved streaks of lilac, brown, and white. The margin white.

The caterpillar is differently said by some to be green with black rings round the body, and by others brown with the segments and a line along the side yellow, and two lines along the back formed alternately of white and red points.

The chrysalis state is said to continue ten days. It is placed in the leaves of plants folded and spun together.

The plate is from Mr. Westwood's figure.

PURPLE EMPEROR.

EMPEROR OF MOROCCO.

PLATE XXXVI.

Apatura Iris,	OCHSENHEIMER. LEACH.
" "	STEPHENS. CURTIS. DUNCAN.
Doxocopa Iris,	HUBNER.
Papilio Iris,	LINNÆUS. LEWIN. WILKES.
" "	HAWORTH. DONOVAN. HARRIS.

THE 19th. of July, 1852, must ever be the most memorable one the events of which are recorded in my entomological diary, for on that day did I first see the Emperor on his throne—the monarch of the forest clothed in his imperial purple,

'Mille trahens varios adverso sole colores.'

"The velvet nap which on his wings doth lie,
The silken downe with which his back is dight,
His glorious colours, and his glistering eies."
SPENSER.

One! two!! three!!! "Allied Sovereigns!" Thanks to the obliging hospitality of the Rev. William Bree, the curate of Polebrook, to whom I had no introduction but that which the freemasonry of entomology supplies to its worthy brotherhood, I had the happiness of beholding His Majesty, or to speak more correctly, Their Majesties, though, as is only proper, at a most respectful distance; they at the "top of the tree," and I on the humble ground. The next day, in the same wood, at Barnwell Weld, near Oundle, Northamptonshire, during my absence in successful search of the Large Blue, of which more anon, Mr. Bree most cleverly captured one, by acting on the principle—an invaluable one, as I have always found it, long before its enunciation by the late Sir Robert Peel, to the students of the University of Glasgow, at

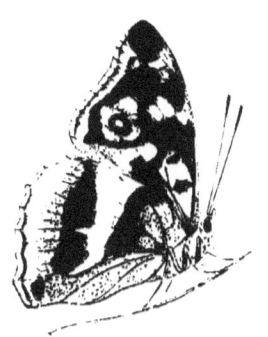

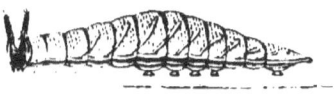

his installation as Rector, in the best speech, by the way, if I do not make my sentence too long, that he ever made—namely, whatever you want to do that is within the bounds of possibility, determine that it shall be done, and you will be sure to succeed! That specimen, a male, as a practical illustration of the lesson, now graces my cabinet, together with the first female that its captor had ever taken, both obligingly presented by him to me. Since then, I have just heard from him that he took another the day after I left him, in one of the ridings of the wood, in his hat. I hope that Her Most Gracious Majesty has no more profoundly loyal subject than myself, and I may therefore relate that, while plotting and planning an "infernal machine" against His Imperial Majesty's liberty and life the following summer, in the shape of a fifty-foot net, and without any reference therefore to what is now going on in France, or any illusion to the career of Louis Napoleon, my toast that evening after dinner was, (with as much sincerity as in the minds of the French,) '*Vive L'Empereur!*' Since then, in 1854, Mr. Bree captured nine in one day in three hours, three of which he has given to me.

The following are given as localities for this noble fly:—The neighbourhood of Doncaster, Yorkshire: but I must frankly confess that I never saw it there; Warwickshire; in Hampshire, the neighbourhood of Winchester, J. Wesley, Esq. has informed me, and also the Isle of Wight; Coombe Wood and Darenth Wood, near London; Bradfield, near Reading, and Enborne Copse, near Newbury, Berkshire; Sywell Wood, near Northampton, Lilford, Barnwell, and Ashton Wold, and the neighbourhood of Polebrook, Northamptonshire; Haslemere Woods, in the neighbourhood of Arundel, and Poynings, near Brighton, Sussex; near St. Neots, Huntingdonshire; Anstey, in Warwickshire. In the woods near Stoke-by-Nayland, Suffolk, R. B. Postans, Esq. tells me that it is found abundantly, as it also is in those of Badley, Dodnash, and Raydon, and he has favoured me with a fine specimen. He captured six in 1851, one of them reared from the caterpillar, and he was informed by Mr. Seaman, an old collector at Ipswich, that in Hartley Wood, near St. Osyth, and between Dedham and Colchester, in Essex, he in one season took a hundred specimens in a fortnight. It is also taken in that county in Epping Forest, Great and Little Stour Woods, Wrabness, and Ramsay; Lyndhurst, in Hampshire; Clapham Park Wood, Bedfordshire; and Brinsop Copse, Herefordshire. Sir Charles Anderson, Bart., saw some by a wood between Grove and Sturton, Nottinghamshire.

This splendid insect is to be seen, if seen at all, the first or second week in July, perched on the outermost spray of some commanding oak or other tree—an elm or an ash—the highest that the neighbouring locality affords him. There he sits, generally with his attention directed outwards, as an Island King's should be, conscious that at home he is secure. If a rival approaches, a fight is of course the consequence—'Pares cum paribus:' and "O 'tis a goodly sight to see!"

The caterpillar is to be found at the end of May.

It feeds on the broad-leaved sallow.

The wings expand to the width of from two inches and three quarters to three and a quarter. The fore wings are of a blackish hue, with a most splendid purple iridescent colour apparent in a proper light. In the middle and towards the outer margin at the tip, are three series of white spots, two, five, and two, the inner ones conjoined forming the waved upper end of a bar which run nearly across the hind wings. In these wings the same splendid purple colour is observable, though scarcely so objectively; in some lights they too are dull "lack-lustre" black. A fulvous line follows the margin, and within its outer corner is a small obscure fulvous spot. Near the lower corner is a fine eye—a black pupil with a light centre and an orange rim—and outside it some fine fulvous marks.

Underneath, the fore wings are varied with silvery greyish white, grey, orange, fulvous, and black, a white band, formed of interrupted spots, running irregularly across them, behind which is a black eye, with a lilac-coloured centre, surrounded by a broad orange circle, in which are two white spots. The hind wings are grey, with a broad silvery greyish white bar across them, tapering towards the corner, with a broad ferruginous adjoining band on each side, but much the least distinct on the inner. The corner is also ferruginous, and above it is a black eyelet, with a lilac-coloured pupil and orange centre.

The wings of the female are of a general blackish brown ground colour, and the markings the same as in the male. The larger of the measurements given above are hers.

The caterpillar is green, with pale yellow oblique lines.

The chrysalis is of a pale green colour.

The plate is from a specimen in my cabinet—the one captured by the Rev. William Bree.

PURPLE HAIRSTREAK.

PLATE XXXVII.

Thecla quercus,	LEACH. STEPHENS. CURTIS.
" "	BOISDUVAL. DUNCAN. WESTWOOD.
Papilio quercus,	LINNÆUS. LEWIN.
" "	DONOVAN. WILKES. HARRIS.
Bithys quercus,	HUBNER.
Lycœna quercus,	OCHSENHEIMER.

THIS is a sort of miniature of the Purple Emperor, though not a "flattering likeness," the wings reflecting something of the same iridescent purple colour, but inferior both in extent and intensity; it is, however, a very pretty insect.

It is common throughout England in most parts of the country. I have met with it at Nunburnholme in the Brant Wood, and at Buttercrambe Moor, and it also occurs at Sutton-on-Derwent Wood, and Raincliff Wood near Scarborough. Sandal Beat near Doncaster, Yorkshire, and in the vicinity of Charmouth, Dorsetshire; also in Sywell Wood, near Northampton, and Lyndhurst, Bisterne, and Shanklin in the Isle of Wight, in Hampshire; in Sussex, near Brighton. Barren Wood near Carlisle, in Cumberland; at Barnwell and Ashton Wold, and in the neighbourhood of Polebrook, Northamptonshire. Near Great Bedwyn and Sarum, Wiltshire, it likewise occurs, but not commonly there. In Scotland it is rare. In Ireland, A. G. More, Esq., of Trinity College, Cambridge, has met with it in plenty at Ardrahan, in the county of Galway. In Wales, near Gloddaeth.

The middle of July is the time for the appearance of the Purple Hairstreak, but it is sometimes still out until the latter end of September.

It is to be seen flying over the tops of oak trees in and near woods.

The caterpillar is found in the beginning of June and in July.

It feeds on the oak.

This fly varies in the expanse of its wings from about an inch and

M

a quarter to an inch and a half. The fore wings are blackish brown all round and over the whole of the upper outside corner, the outside edge white, the remainder being filled in with iridescent purple, more or less extensive and intense. The hind wings, which have a short tail, are uniform blackish brown; the outside edge is white.

Underneath, the front wings are bronzed ash-colour, with a slender white streak shadowed on the inside with brown some distance within the outer margin, and following it, but not all the way; near the lower corner are two orange fulvous marks, the base of a faint procession of the same upwards, obscurely visible in some lights. The hind wings are of the same bronzed ash-colour, with a similar white streak, taking up that of the fore wings, but slightly curved inwards instead of following the margin; near the lower corner are two other orange fulvous spots, the upper one with a black centre, and the outer one edged at its base with black, which runs on in a streak to the tail. The antennæ are reddish on the lower side of the clubs.

The female resembles the male, but the purple is obscurely extended over the black of the upper wings in some lights, and likewise over the lower ones, excepting a rather broad band at the margin.

The caterpillar, which in appearance somewhat resembles a woodlouse, is short and thick, of a dull rose-red colour, covered with short hairs, and with several rows of dark greenish lines or dots. It sometimes, when full fed, goes under the ground.

The chrysalis is described as being of a shining rusty brown colour, with three rows of brown spots on the back.

GREEN HAIRSTREAK.

PLATE XXXVIII.

Thecla rubi,	LEACH. STEPHENS.
" "	CURTIS. DUNCAN.
Papilio rubi,	LINNÆUS. LEWIN. HAWORTH.
" "	DONOVAN. WILKES. HARRIS. ALBIN.
Lycus rubi,	HUBNER.

This "petite" species is not uncommon, though only of local distribution. I have taken it in tolerable plenty at Buttercrambe Moor, near Stamford Bridge, Langwith, Stockton, and Sand Hutton, Yorkshire, and the warren on the east cliff near Charmouth, Dorsetshire. In the following places it is also to be found:—Barnwell and Ashton Wold, and the neighbourhood of Polebrook, Northamptonshire; near Great Bedwyn and Sarum, Wiltshire; and near Winchester and Bisterne, in Hampshire; Brighton, in Sussex; Carlisle, Cumberland. In Wales, near Marle and Conway. It occurs throughout the whole of England in suitable situations, but in Scotland only in the southern districts. It is abundant in the Isle of Mull.

It is attached to woods and waste covers, and gardens near these.

This pretty little insect appears in the perfect state at the end of May or beginning of June, and a second brood appears the beginning of August. It frequents thorn and bramble bushes in the more uncultivated parts of the country.

The caterpillar is to be found in June and the middle of July.

It feeds on the bramble, *(Rubus fruticosus,)* broom, *(Spartium scoparium,)* dyer's weed, *(Genista tinctoria,)* and other plants.

The wings expand to the width of from rather over an inch to an inch and a quarter. The fore wings are of a uniform bronzed brown colour, with a dark spot in the middle near the front edge. The base of these wings has a tinge of green. The hind wings are of a similar colour.

Underneath, the fore wings are of a beautiful metallic green colour, except at their lower part, which is brown. The hind wings are also of the same green colour, with a row of minute white dots, forming a more or less complete streak near the middle, but outside it, following the shape of the margin. In some specimens the dots are almost obsolete. The tails are very short, and hardly distinguishable from the other projections which border these wings.

Mr. Stephens describes a variety in which the fore wings have a row of white dots on their front margin.

The female resembles the male.

The caterpillar is light green, with rows of triangular yellow spots on the sides: the head is black.

WHITE-W HAIRSTREAK.

PLATE XXXIX.

Thecla W-album,	HUBNER. GODART. STEPHENS.
" "	CURTIS. DUNCAN.
Papilio W-album,	VILLARS.
" *pruni,*	LEWIN. HAWORTH. DONOVAN.
Thecla pruni,	LEACH. JERMYN.
Strymon W-album,	HUBNER.
Lycæna W-album,	OCHSENHEIMER.

HAPPY in all its aspects is "Rural life in England," and not the least so when the love of Natural History gives a zest to every walk and ride, and invites you moreover into calm and peaceful scenes, where, for the most part, the beauties of nature are seen to the greatest advantage.

I have captured this pretty species in numbers at High Melton Wood, near Doncaster, and Nunburnholme Brant Wood. Edlington Wood is another locality. It is rather difficult to find it at a sufficiently low elevation. It has likewise been obtained at Barnwell and Ashton Wood, and the neighbourhood of Polebrook, Northamptonshire; near Windsor, in Berkshire; Ipswich, Playford, and Bungay, in Suffolk; Allesley, in Warwickshire; Southgate, in Middlesex, also at Chertsey; and in Cambridgeshire; so too, in the vicinity of Ripley, in Surrey, where it was observed by Mr. Stephens, in myriads in the year 1827. It is partial to the bramble blossom.

The fly occurs in July, also in September.

The caterpillar is to be found the beginning of June.

It is said to feed on the elm and the blackthorn.

The wings expand to the width of from a little under an inch and a quarter to nearly an inch and a half. The fore wings are of a uniform dark blackish brown. The hind wings are of the same dark ground-colour, with a minute white dot outside the tail, from which a very narrow white streak runs to and round the tail; there are a few rufous marks near the lower inside corner.

Underneath, their general colour is a fine rich ash brown, with a white line nearly across them, and then fainter, turning inwards. The hind wings have also a slender white line across them beyond the middle, thinner still towards the lower side, where it bends in a zigzag shape, forming the letter W. A row of small black crescents, slightly edged on the inside with white, runs nearly parallel to the outer margin of these wings, and is succeeded by a fulvous band extending from the inner lower corner, about half-way towards the outer angle, when it becomes gradually obliterated; on its outside this band is marked with black semicircular spots, succeeded by a faint silvery line, the black spots nearest to the lower corner being the largest; the lower corner itself is black, with a silvery dot. The tails are black, tipped with white; the antennæ are ringed very prettily with white, the tips red; the legs, whitish, ringed with brown.

The female has the white streak on the fore wings rather broader than in the male, as also rather more waved, and they are without the spot on the middle of these wings; the tails to the hind wings are a little longer than in the male.

The caterpillar is pale yellowish green, covered with short downy hairs, with the two rows of small dots down the middle of the back, which is indented, and paler oblique marks or lines on the sides; the hinder segments are spotted with dark red.

The figures are from specimens in my own collection.

BLACK HAIRSTREAK.

PLATE XL.

Thecla pruni,	STEPHENS. CURTIS. DUNCAN.
" "	WESTWOOD.
Papilio pruni,	LINNÆUS. HUBNER.
Strymon pruni,	HUBNER.
Lycæna pruni,	OCHSENHEIMER.

BARNWELL and Ashton Wold, and the neighbourhood of Polebrook, Northamptonshire, where the Rev. William Bree has captured it in plenty, Black Hill, near Exmouth, Devonshire, where James Dalton, Esq., of Worcester College, Oxford, has taken it, and Monk's Wood, in Huntingdonshire, are localities for this fly.

It appears the middle of July, but has been taken I believe so soon as the 18th. of June.

This species averages about an inch and a quarter in the expansion of its wings. The fore ones are of a blackish brown colour, with a light silky patch near the middle towards the front edge. The hind wings have two or three, or more, pale orange-coloured semicircular-shaped spots near their hind margin at the inner corner, and extending more or less forwards, often running much together into a line.

Underneath, the ground-colour is an ash grey, the fore wings having a slender, nearly straight, bluish white line extending nearly across them beyond the middle; outside this line there are several obscure fulvous patches, those nearest to the lowest corner being preceded by a black and silver dot or eyelet. The white line proceeds from the fore wings, and reaches to the inner margin of the hind wings, where it becomes more irregular, somewhat resembling an obtuse W. The black spots, seven in number, and edged internally with silver, are more conspicuous on these wings, and are succeeded by a broad fulvous bar extending to the outside corner, its outer edge being marked with semicircular black marks, followed by a silvery line. The spots nearest the lower inside corner are the largest, and the corner itself is black with a silvery dot; the tails are black; the antennæ ringed with white, as are the eyes.

The caterpillar is green, with oblique yellowish lines on the sides, and darker marks down the back.

The chrysalis is brown, with lighter markings, and dark tubercles.

Slight varieties occur in the extent of the orange-coloured marks on the upper wings of this insect, but still they are not of sufficient importance but that the rule 'Ex uno disce omnes' may be readily applied to any individual specimen.

BROWN HAIRSTREAK.

PLATE XLI.

Thecla betulæ,	FABRICIUS. LEACH. STEPHENS.
" "	CURTIS. DUNCAN. WESTWOOD.
Papilio betulæ,	LINNÆUS. HAWORTH. DONOVAN.
" "	ALBIN. WILKES. HARRIS. LEWIN.
Lycæna betulæ,	OCHSENHEIMER
Strymon betulæ,	HUBNER.

How well one remembers the "Long time ago," with which so trivial a thing as the capture of an insect, even though of no great rarity, is associated. The Brown Hairstreak I first, and indeed for the only time, captured, on one of two little hills with an unpleasing designation in the neighbourhood of Wallingford, Berkshire. Barnwell, Ashton Wold, and the neighbourhood of Polebrook, Northamptonshire; near Great Bedwyn and Sarum, Wiltshire; Coombe Wood, Birch Wood, Darenth Wood, and Hornsey Wood, near London; Henfield, in Essex; Raydon Wood, near Ipswich, Suffolk; Dartmoor, in Devonshire; and places in Dorsetshire and Norfolk are also given as localities for it. It has also occurred near Winchester in Hampshire, Brighton in Sussex, at Grange in Lancashire, and in plenty in the valley of the Dovey between Machynlleth and Cemmaes; also near Cork.

It is by no means a plentiful species, though widely distributed.

About the end of August or the first week in September the Brown Hairstreak is to be taken flying about oak and beech trees. It has occurred in July.

The caterpillar occurs at the end of May.

It feeds on the blackthorn, birch, plum, etc.

In this species the wings extend in width from one inch and a third to rather more than one and a half. They are of a rich glossy brown colour, with a short oblong mark near the middle of the front, outside which is a more or less visible lighter mark, but sometimes it is entirely wanting. The outside margin is elegantly bordered with white.

The hind wings have a round orange dot at their lower inside corner, and another at the base of the tail.

Underneath, the ground-colour is fine orange yellow, the edge with a brighter border; the dark oblong mark shews through, and outside it is a long narrow wedge-shaped mark, its base arising at the front edge and running two thirds across the wing. It is edged all round with a very slender dusky line, and this again on the outside by a narrow white one. The hind wings are of a still richer brown colour, especially at the lower inside corner: a darker orange wedge-shaped bar, edged with a narrow dusky line, and this by another narrow white one, runs almost entirely across them, its wide end beginning at the middle of the front upper edge. The tail is bordered with a black edge, which proceeds to the inner corner.

The female is of the same glossy brown colour, and has a large orange brown patch between the middle and the outside corner, and running nearly across them, gradually narrowing; at the upper corner of the patch is a dark oblong mark, as in the male. The hind wings have the like orange dot, and the mark on the tail, but rather more extensive. Underneath, she resembles the male.

The caterpillar is light green, with paler oblique lines on the sides, and straight ones down the back.

The chrysalis is brown, with darker marks.

DUKE OF BURGUNDY FRITILLARY.

PLATE XLII.

Nemeobius Lucina,	STEPHENS. HORSFIELD. DUNCAN.
" "	BOISDUVAL. WESTWOOD.
Hamearis Lucina,	HUBNER. CURTIS. WESTWOOD.
Papilio Lucina,	LINNÆUS. LEWIN. DONOVAN.
" "	HARRIS. HAWORTH.
Melitæa Lucina,	OCHSENHEIMER. LEACH. JERMYN.

"PARVUM parva decent," says the proverb, but the high-sounding and sesquipedalian name of this small species is by no means in harmony with its diminutive size. It is not however my province to write a work on "Titles of Honour," nor to give any genealogical account of the Duke of Burgundy Fritillary. So far nevertheless the name is appropriate, in that Dukes and these butterflies are alike somewhat rare, and from my blazon of the plate it will be seen that the latter, as is only Ducal, have numerous quarterings.

I have taken this pretty insect in tolerable plenty in the neighbourhood of Melton Wood, near Sandbeck Park, Tickhill, Yorkshire. It occurs also in Sywell Wood near Northampton, and at Barnwell and Ashton Wold, and in the neighbourhood of Polebrook; and, though rather uncommon there, near Great Bedwyn and Sarum, Wiltshire; Coombe Wood, near London; Darenth Wood, Kent; Boxhill and Dulwich; the New Forest, in Hampshire; Brighton, in Sussex; Bagley Wood, Oxfordshire; and in Dorsetshire and Berkshire. Mr. Heysham has taken it as far north as Carlisle. It is not uncommon on the banks of the Eden. Also in Wales, near Llandudno.

It is out the beginning of June.

The caterpillar is stated by Hubner to be found before midsummer, after which period it turns into the chrysalis state.

It feeds on the common primrose, (*Primula veris,*) and the broad-leaved primrose, (*Primula elatior.*)

This species varies in the expanse of its wings from a little under to a little over an inch and a quarter. On the upper side the fore wings are dark fulvous, crossed with three waved and indented bars of dark blackish brown, the inner one being the most irregular, the indentations of each meeting the next one, forming a kind of mosaic work; the base is entirely of the dark blackish brown colour; the extreme margin is yellowish, intersected at intervals by the outside bar, which is close to it, and immediately inside which is a row of dots in the fulvous part, which is a row of crescents formed by the extensions of the next dark bar. The hind wings are nearly entirely of the dark colour, the edge being a row of widely interrupted whitish yellow marks, inside which is a row of fulvous crescents, each with a black triangular-shaped dot within it, and in the centre of the wing three fulvous dots, semicircularly disposed, between which and the base is one other, which may indeed be considered as the upper part of the curve in the shape of a sickle.

Underneath, the fore wings are of a paler fulvous ground-colour, the outside edge being cream-colour, indented with brown, the centre of the wing being paler than the rest, with two irregular short interrupted rows of large dark brown marks on it, and towards the inner edge and the lower corner, and towards the tip a row of light dots partly following the margin, with one extra one between the row and the tip itself. The hind wings are of a rather darker and richer fulvous, crossed with two irregular waved bars formed of large silvery cream white spots, one near the base and the other about the middle; the latter has some dark markings on its inner edges, and a few more between it and the outside, on the fulvous ground-colour.

In the female there is a greater extension of the pale colour on the upper surface of the fore wings, and the blackish brown colour is darker.

The eggs are found solitary or in pairs, on the under side of the leaves of the primrose. They are round, smooth, shining, and of a pale yellowish green colour.

The caterpillar is of an oval but depressed and elongated shape; the head rounded, heart-shaped, smooth, shining, and of a bright ferruginous colour; the body is covered with rows of tubercles: it is set with feathery hairs. On the hinder part of the back there is a black dot on each joint, and on the sides the like, but the spots less distinct. The general colour is pale olive orange; underneath, it is whitish; the feet are rusty brown; the claws whitish. It moves very slowly, rolls itself up when disturbed, and remains in that state a long time. Hubner says that it changes its skin five times before going into the chrysalis state, and that each appearance varies considerably.

The chrysalis is suspended from the head, and is also kept by a cord round its middle.

The figure is from one of the specimens in my own cabinet.

GREASY FRITILLARY.

MARSH FRITILLARY.

PLATE XLIII.

Melitæa Artemis,	FABRICIUS. OCHSENHEIMER. HUBNER.
" "	STEPHENS. DUNCAN. WESTWOOD.
Papilio Artemis,	FABRICIUS. LEWIN. HARRIS.
" *Maturna*,	ESPER.
" *Lyc*,	BORKHAUSEN.
" *matutina*,	THUNBERG.
" *Lucina*,	WILKES.

THE following are localities for this fly:—Finstall, near Broomsgrove, Worcestershire, where, as in other places in that neighbourhood, for instance, outside the coppice beyond Grafton Manor House, and in the low fields near Whiteford Mill, I have taken it; Barnwell and Ashton Wold, Aldwinkle, the birth-place of Dryden, and the neighbourhood of Polebrook, Northamptonshire; Great Bedwyn and Sarum, Wiltshire; Durdham, near Bristol; Eriswell, near Beccles, and Mildenhall, Suffolk; Coleshill and Coventry, Warwickshire; Woodstock, Oxfordshire; Glanville's Wootton, Dorsetshire; Dartmoor, Devonshire; Enborne Copse near Newbury, Berkshire; Holme Fen and Monk's Wood, Huntingdonshire; Clapham Park, Bedfordshire; Brighton, Sussex; Beachamwell, Norfolk; Stockton Common near York, Buttercrambe Moore near Stamford Bridge, and Langwith; Belford, Northumberland; Haverfordwest, Pembrokeshire; and A. G. More, Esq., of Trinity College, Cambridge, has seen it in plenty near Ardrahan, in the county of Galway, Ireland.

It is found in marshy meadows and marshy places in woods, as also on heaths in some places.

The perfect insect appears the middle of May, and continues into June and July.

The caterpillar, which is hatched in the autumn, in August, and is gregarious, the brood passing the winter under a common web, is full fed in April.

It feeds on the devils-bit scabious, *(Scabiosa succisa,)* the greater plantain, *(Plantago major,)* and the ribwort plantain, *(Plantago lanceolata.)*

This Fritillary extends across the wings from one inch and a half to two inches: I have two in my collection of exactly these respective measurements. The fore wings are of a dark reddish orange colour, barred cross-wise irregularly with blackish and straw-coloured waved bars or spots; the base of these wings is blackish brown. The hind wings are of a similar red ground-colour, their base and inner side blackish brown, with a yellowish orange spot near the former; they are also barred with a bar of dark blackish brown, widest at the lower corner, and in it a row of continuous light orange cream-coloured spots; the ground-colour, there apparent as another bar, has a row of black specks in each of its compartments, and this is succeeded by a blackish brown border, edged with yellowish grey, and bordered on its inner side with a row of small yellowish orange crescents, each a satellite of the several compartments of the red ground-colour.

Underneath, the fore wings, which have an oiled appearance, whence the common name of the species, are of a much more obscure and dull general colour, the markings from the upper side all, or nearly all, shewing through. The hind wings, of a slightly brighter ground-colour, have three yellowish cream-coloured curved bands, margined with thin black lines, the first, near the base, irregular and oblique, with an extension outwardly into the middle, the second, in the middle of the wing, and the third, a series of marginal crescents, between which and the middle one, in the ground-colour, is a row of small yellowish cream-coloured spots, with a central dot of black, severally more or less distinct.

The female resembles the male.

The caterpillar is spined, black above and yellowish beneath, with a row of small white dots along the back and sides; the spines are black, as is also the head; the legs are reddish brown.

The chrysalis is suspended, according to M. Harris, between several blades of grass, drawn together, and fastened at the top with threads. It is pale-coloured, with dark spots. It continues about a fortnight in its "durance vile," and then the beautiful insect emerges to the full enjoyment of its "little day."

Individuals vary considerably in the intensity and size of the markings. In one, the front edge of the fore wings was slightly concave.

The figures are from specimens in my own collection.

GLANVILLE FRITILLARY.

PLATE XLIV.

Melitœa Cinxia,	OCHSENHEIMER. BOISDUVAL.
" "	STEPHENS. CURTIS. WESTWOOD.
Papilio Cinxia,	LINNÆUS. LEWIN. DONOVAN.
" "	WILKES. HARRIS.
" *Delia*,	HUBNER.
" *Pilosellæ*,	ESPER.
" *Trivia*,	SCHRANK.
" *Abbacus*,	RETZIUS.

THIS butterfly is a very local one, so that its capture must always be regarded as a "great fact" in the experience of by far the greater number of entomologists.

J. W. Lukis, Esq. informs me that this extremely interesting insect is taken, though very rarely, in the neighbourhood of Great Bedwyn and Sarum, Wiltshire. It seems to be most plentiful near Ryde and other places in the Isle of Wight, on the grassy sides of the little glens which run down to the sea-shore, and on the edges of the cliffs, where I have seen it in goodly numbers myself. One was captured by Mr. Walhouse near Leamington, in Warwickshire; Dover, Dartford, and Birch Wood, in Kent, are likewise given as localities for it: it is said also to have occurred in Yorkshire and Lincolnshire.

The Glanville Fritillary appears the end of May, and in June and July.

"The caterpillars are found," says Mr. Westwood, "in the autumn, living in societies under a kind of tent formed by drawing together the tips of the leaves on which they feed, and covering them with a web."

This butterfly varies in the expanse of its wings from a little under to a little over one inch and three quarters. The fore wings are of a rich fulvous ground-colour, elegantly tesselated all over the outside half with black brown markings, arranged for the most part in waved cross lines, intersected by longitudinal ones; the margin is yellowish cream-colour, indented with the ends of the last-named lines; the base of the wings is blackish brown, as is the lower edge; the black markings on the inner portion of the wings are hollow, the fulvous-colour apparent on their centres. The hind wings are marked in a very similar manner, with the addition of a row of dots inside the last black bar, between it and the next one, in the middle of each fulvous intersection.

Underneath, the fore wings are rather lighter fulvous, the corner only of the transverse bars shewing through, and the inner markings faintly; the whole of the tip is broadly washed with pale dull yellow, with a waved black line and a row of black dots, which skirt the edge of the outside of the wings, succeeded by a straw-coloured margin, intersected with black. The hind wings have three broad straw-coloured bars across them, the inner one at the base, and the outer on the outside part; the inner one with a row of black dots along its middle, the middle one with a row of the like at its inner edge, and the outer one with a waved line of black; the two intermediate spaces are pale fulvous, the inner of them with a straw-coloured patch in its centre, and the outer one with darker marks of fulvous in its central divisions, which are marked out with thin transverse lines.

The caterpillar is of an intense black colour, very slightly spotted with white; the head and the fore wing fulvous.

The chrysalis is brownish, with rows of raised fulvous marks on the back.

The engraving is from specimens in my own cabinet.

PEARL-BORDERED FRITILLARY.

PLATE XLV.

Melitæa Euphrosyne,	LEACH. STEPHENS. CURTIS.
" "	DUNCAN. WESTWOOD.
Papilio Euphrosyne,	LINNÆUS. LEWIN. DONOVAN.
Argynnis Euphrosyne.	OCHSENHEIMER. HUBNER. BOISDUVAL.

To see this pretty insect gaily flitting about in the open places in woods in the new summer time, when every trace of winter has at last disappeared, is almost enough to make one wish to be, as well as to sing "I'd be a butterfly," so happy and joyous does it seem.

It is very plentiful in many places; among others, at Buttercrambe Moor, near Stamford-Bridge, Stockton, Sutton-on-Derwent, and Langwith, Yorkshire; near Great Bedwyn and Sarum, Wiltshire; and in great abundance in all the woods near Shelly, Stoke-by-Nayland, Suffolk; Hainault Forest, in Essex; Brighton, Sussex; Birch Wood, in Kent; Bisterne, in Hampshire; and Barren Wood, near Carlisle It is also abundant in various parts of Scotland; also in Wales, as near Llandudno.

It is found in woods.

There are two broods of this butterfly, the first appearing the end of May and beginning of June, and the second in August and September.

The caterpillar feeds on the wild violet, *(Viola canina,)* and other species of that genus.

In this insect, which measures from one inch and three quarters to nearly two inches across, the upper side is fulvous, mottled over with several large black billets on the centre of the wings, placed in a zigzag manner, the inner series running across the wing in a connected manner, followed by two other sets, which only extend half-way across from the front edge; the base of the wings is blackish brown, much more extensive and distinct in some individuals than in others. Next the outside series is a row of round dots, succeeded by a row of small crescents, and these by a row of round dots intersected by a line and forming the margin, the brown of the wings appearing between the circles. The hind wings are marked in a very similar manner, but the billets are much run together. Their base is also dusky black.

Underneath, the fore wings exhibit all the markings from the upper side, the tips being lighter, and bordered some way within with dark reddish orange. The hind wings have one large silver spot on their centre, placed diagonally across the central pale bar, and a row of silver crescents runs round the edge of the wing, followed by a reddish line, itself margined by a streak of greenish straw-colour, the outside margin. These wings are very beautifully marked with reddish ferruginous, buff, and greenish straw-colour, in the way of waved bars, formed by a wide band of the former mottled with the latter, this by one of the greenish straw-colour, and this by one of the latter, the base being also of the greenish straw-colour; the central pale bar has one.

The caterpillar is black in colour and spined, with two rows of orange dots on the back.

Several varieties of this species have been recorded.—The Rev. C. J. Bird, Vicar of Gainsborough, possesses one which is nearly white. Mr. Stephens recorded one in which the silvery marginal spots are wanting, and another with the inner half of all the upper surface of the wings black, spotted with fulvous, with large black spots on the under side of the fore wings. Mr. Westwood also figures one in which all the black markings on the upper side of the fore wings are suffused, except the row of round spots within the margin, the markings on the hind wings being somewhat more distinct, and the under side scarcely different from the ordinary appearance. The autumnal brood is of a much yellower ground-colour than the spring one.

The figures are from specimens in my own collection.

SMALL PEARL-BORDERED FRITILLARY.

APRIL FRITILLARY.

PLATE XLVI.

Melitæa Selene,	STEPHENS. CURTIS. DUNCAN.
" "	WESTWOOD.
Papilio Selene,	FABRICIUS.
" *Silene,*	HAWORTH.
" *Euphrasia,*	LEWIN.
" *Euphrosyne,* (var:)	ESPER. HARRIS.

THIS is another of those insects which do not make their appearance until the cold weather has fully passed away, and there is no longer any danger, as is the case with other less fortunate species, of mistaking some transient gleam of sunshine for the established serenity of the true summer.

It frequents woods, heaths, and waste grounds. I have taken it in plenty in Edlinton Wood, near Doncaster, an excellent locality, and one in which nightingales abound; also at Buttercrambe Moor, Langwith, and Stockton. Other localities are Barnwell, and Ashton Wold, and the neighbourhood of Polebrook, Northamptonshire. It is scarce near Falmouth. It occurs also near Great Bedwyn and Sarum, Wiltshire; Raydon Wood and Layer Heath, near Colchester, Essex; Dartmoor, in Devonshire; Lyndhurst, and Bisterne, in Hampshire; Newcastle, in Northumberland; Ambleside, in Westmoreland; near Carlisle, in Cumberland; Looe, in Cornwall; and Durham, "of that ilk." Brighton, in Sussex; and Llandudno, in Wales.

There are two broods in the year, the one appearing in May and June, and the second in August and September.

This insect measures in the expanse of its wings from a little under to a very little over an inch and three quarters. The ground-colour of the fore wings is fulvous, the base black, with a series of waved lines of irregular black dots of different sizes and shapes, and outside this, from one larger black spot near the tip, two rows of six dots, the inner larger and rounder, and the outer smaller and of a triangular shape, which run across the wing; the margin is similarly marked with black dots on a narrow black line, running all round the outside and the base of these wings. The hind wings are marked in a very similar manner, and their base is also black, running into the wing, and forming the inside bordering of the central markings.

Underneath, the fore wings are much the same as on the upper side, but the ground-colour is more dull, and the black marks, which shew through, are not so large; the tip is of a paler hue than the rest of the wings, and it is divided by a reddish ferruginous waved bar, running out below to the edge in a loop. The hind wings are very elegantly marked with reddish ferruginous, buff, and greenish straw-colour, in the way of waved bars, formed by a wide band of the former mottled with the latter, this by one of the greenish straw-colour, and this by one of the latter, the base being also of the greenish straw-colour; the central pale bar has one large spot of silver placed diagonally across it, of an oblong, quadrangular, uneven shape, another inside it of an oblong form, placed upright, wedge-shaped at each end, and another divided in three by the veins of the wing running from near the base of the former to the upper edge of the wing, near the outside corner; between these three patches and the lower corner is another horizontal one, divided by the red, near its outer and smaller portion, and a row of crescents runs round the edge of the wing, the middle ones silver, followed by a thin reddish line, itself margined by a streak of the greenish straw-colour, which forms the outside margin.

The caterpillar is black, with a pale stripe along the sides. The spines are half yellow, and two on the neck are longer than the others, and project forward.

The chrysalis is of a dull grey colour.

This species is liable to vary considerably. Mr. Stephens describes one specimen in which the upper surface of the wings was whitish. Another, recorded as a separate species by the name of Thalia, is described as having the wings above pale fulvous, irregularly spotted with black; the front ones, underneath, pale, varied with yellowish and ferruginous towards the tips, with some obsolete black and dusky spots; the hind wings variegated with ferruginous, yellowish, and greenish, with the pupil of the eye very large; the silvery spot continued to the hind margin, and the usual marginal spots lengthened inwardly, the usual bands obliterated, but the silvery spot at the base somewhat apparent.

The figures are from specimens in my own collection.

PEARL-BORDERED LIKENESS FRITILLARY.

WHITE MAY FRITILLARY. HEATH FRITILLARY.

PLATE XLVII.

Melitæa Athalia,	OCHSENHEIMER. STEPHENS. DUNCAN.
Papilio Athalia,	ESPER.
" *Dictynna,*	LEWIN. HAWORTH. JERMYN.
" *Maturna,*	FABRICIUS. WILKES. HARRIS.
Cinclidia Athalia,	HUBNER.

THE late Captain Blomer used to take this very interesting insect plentifully in Devonshire; Ford Wood is one of the localities there for it, and Dartmoor another; it is taken also in Cain Wood, Middlesex: Bagley Wood, Berkshire, near Oxford; Apsley Wood, and near Bedford, Bedfordshire; near Deal, Faversham, Canterbury, and at Combe Wood, Kent. It occurs near Falmouth, in Cornwall, but rarely; W. P. Cocks, Esq. has taken it there, likewise at Looe; also not very uncommonly near Great Bedwyn and Sarum, Wiltshire, as J. W. Lukis, Esq. informs me; and at Langham Lodge Wood plentifully, as too in Hartley Wood and Maldon Wood, near St. Osyth, and in the High Woods, near Colchester, Essex; it has also been taken at Peckham, Surrey, near London. I have never yet seen it on the wing, but live in hope of being able to chronicle its capture on some "Red-letter day."

It is "out" in the end of June and the beginning of July; the end, and even the beginning of May, is also given as a date for it; but I think it must be a mistake. It frequents woods, heaths, and marshes.

The caterpillar feeds on the broad-leaved plantain, (*Plantago major,*) and the narrow-leaved plantain, (*Plantago lanceolata.*)

This species is a little over an inch and a half, and from that to three quarters, in the expanse of its wings. The fore wings are fulvous, blackish brown at the base, and waved all over with blackish brown lines, intersected by others running through to the border, which is blackish brown, edged with yellowish white, and indented by the blackish colour. The hind wings are very similar to the fore ones in their markings in all respects.

Underneath, the fore wings are paler fulvous, with a few slight blackish brown marks, indicating the situation of some of the principal markings on the upper side; the tip is straw-colour, which runs some way along the margin, crescented on the inside and across by black lines, crossed by a waved longitudinal line following the edge, and intersected by others running into it, and reaching to the border. The hind wings are very elegantly marked; a whitish cream-coloured bar runs across them, crossed and margined with blackish brown lines; the base is reddish brown, but is very much mottled over with a continuation of the whitish cream-colour; the bar is succeeded by another narrow waved bar of fulvous crescents, margined with black, and outside this is a darker cream-coloured scolloped bar, tinged with yellowish, and edged round each division with blackish brown, the lines running through to the edge, which is also cream yellow, divided from the last-named bar by two lines, one straight and the other crescented.

The caterpillar is spined, and black, with two white dotted lines on each segment, and white tubercles on the sides.

One variety of this insect, described by several authors as a distinct species, under the name of Melitæa Pyronia, a specimen having been taken by Mr. Howard, at Peckham, in Surrey, in June, 1803, is described by Mr. Westwood as rather more than an inch and a half in expanse, with the fore wings above deep fulvous; the veins, blots in the middle, a waved streak, and the marginal bar, black; the hind wings above, black, with a waved bar of six fulvous spots beyond the middle; beneath, the fore wings are fulvous, but paler at the tips, with two black spots at the base, and a broad black bar in the middle, divided by fulvous veins, and with a row of black lunules near the margin; the hind wings fulvous at the base, with about eight confluent black patches; the middle of the wing occupied by a brown whitish band, intersected by blackish veins, followed by a row of fulvous lunules, with black edgings; the outer margin straw-coloured, with a row of ochraceous lunules in the middle.

Another variety also described as a distinct species, under the name of Papilio tessclata, is mentioned as being paler than the ordinary colour, and the fore wings more fulvous beneath; the hind wings, underneath, entirely straw-coloured, with black veins; at the base three large yellow spots, edged with black; a broad curved band of straw-colour, edged with black, and with an irregular black line running through the middle of it, across the centre of the wings, and this succeeded by a row of black crescents, the margin being straw yellow, with a black vandyked line running along it.

The figures are from specimens in my own collection.

WEAVER'S FRITILLARY.

PURPLE UNDERWING FRITILLARY.

PLATE XLVIII.

Melitæa Dia,	Stephens. Jermyn. Westwood.
Papilio Dia,	Linnæus. Stewart. Turton.
Argynnis Dia,	Ochsenheimer. Hubner.

Mr. Richard Weaver has taken this rare fly at Sutton Park, near Tamworth, and Mr. Stanley near Alderley, in Cheshire.

Certain "Malignants" having doubted the former captures, I feel constrained to rescue an honest man's character from the undeserved imputation. In a letter I received from the late Dr. Shirley Palmer, of Tamworth, Warwickshire, dated so recently as the 23rd. of October, 1852, he says, "I know not whether you are personally acquainted with that extraordinary man: he possesses the most correct eye for the discrimination of species, of any individual whom I have hitherto met with. On several occasions the poor fellow has experienced rather shabby treatment from the entomologists of London and Paris, and I have had to vindicate him from charges of unblushing falsehood and gross negligence, of which I know him to be utterly incapable. His assertion respecting the capture of any rare insect, if made by himself, may be most implicitly relied on." I formerly, when at Broomsgrove school, knew Mr. Weaver personally myself, as a most successful, because a most indefatigable collector, and the opinion of such a man as the late lamented Dr. Palmer he may well be contented with, should this record of it meet his eye. 'Satis est equites plaudere.'

The caterpillar feeds on the sweet-scented violet, (*Viola odorata*,) and there are two broods in the year.

This species measures a little over an inch and a half in the expanse of its wings. The fore wings are of a reddish brown colour, much marked all over with black marks, several near the centre, the base blackish brown, and a row near the outside edge, followed by a

line, which is again succeeded by another line, through both of which a series of oblong marks cross to the margin, which is yellowish white. The hinder wings are also reddish brown, the base being nearly black, and the marks disposed much in the same way as on the fore wings, except that the inner line is waved from each cross mark to the other.

Underneath, the fore wings shew the central dark marks through, and a few of the others; the upper corner and part of the outer edge is paler than the rest. The hind wings are of a brownish purple colour, relieved with darker markings of the same; there are six or seven small silvery spots, intermingled with minute yellowish dots, at the base, a broad irregular band of the same nearly across the wings, then a purple white streak, in which is a series of circular spots, indistinctly eyed, and the margins also silver in indentations.

The caterpillar is described as being black, with the spines white and reddish, the back greyish, with a line along it.

The plate is taken from Mr. Westwood's figures.

HIGH-BROWN FRITILLARY.

PLATE XLIX.

Argynnis Adippe,	Fabricius. Duncan.
" "	Ochsenheimer. Stephens.
Papilio Adippe,	Linnæus. Harris.
" "	Lewin. Donovan.
Acidalia Adippe,	Hubner.

It has been well observed, that all the best and highest enjoyments of man are those which, coming as they do direct from the bounteous hand of the Omnipotent Himself, are not purchaseable with money or any other human commodity. Every aspect under which nature is viewed throws light upon this remark, and gilds it with the unmistakeable lustre of truth. Without philosophising on the connection between mind and matter, upon an investigation into which, if we enter, the result at its close will be that we shall find ourselves on the threshhold at which we first stood, it is certain that mental enjoyments are those of primary moment, and such as convey the highest satisfaction to the intellectual man, to that part of man which even the heathen philosopher perceived to be the man himself. Where can you buy the feelings of the astronomer, or in what Australia discover a reward for the successful mathematician? Where is the El Dorado that will furnish a value which the gifted mechanist would exchange for his sense of the widely extended benefits his fellow-men derive from the years of his patient endeavours? Under what "Mountain of light" shall you discover a hidden treasure, the finding of which shall equal the exalted delight with which you will gaze from its summit on all that lies between it and the surrounding horizon? What sublunary pleasure can realize the depth of feeling with which you gaze up into the vault of the firmament of Heaven? Was Solomon, "in all his glory," arrayed like one of the "lilies of the field?" What pearls do the "dark unfathomed caves" of the ocean conceal that can compare with the view of the ocean itself? What sordid motive can furnish the thrill of devoted loyalty? What price will purchase peace

of mind? What worldly feeling can equal the aspirations of piety, the "Breathings of the devout soul?" What picture can depict the real landscape to the eye of the mind? "Who can paint like nature?" and where is the artist but must borrow both his ideas and his hues from her? What elaborate perfume can vie with the scent of the primrose, the violet, the hawthorn, and the rose? What artificial draught can give the refreshment that "gentle sleep" will bestow even on the "ship-boy" rocked upon the mast? "He giveth His beloved sleep," and who can sell, or who can buy it, if He denies the inestimable boon to the sick or the wearied in body or in mind?

This may be moralising, but moralising should never be out of place, and I wish for readers who can moralise with a "Country Parson," and share with him in the devout feelings which it is his duty to spread as widely as he can. And what is true of nature, the work of God, in any one particular, is true, in this respect, of all; and if it be the right and the good way to find "sermons in stones and good in every thing," let the entomologist be allowed his share in the laudable feeling, and admire, in the elegant butterfly before us, the handiwork of the Almighty and Adorable Creator.

This fine fly is not uncommon in most of the southern counties, among others in Hampshire, near Winchester; also at Bisterne, and, too, at Lyndhurst. In Yorkshire, at Sutton-on-Derwent Wood and Buttercrambe Moor. It is taken in plenty so far north as Osberton, in Nottinghamshire, the seat of George Savile Foljambe, Esq. Also at Ashton Wold, Northamptonshire, and south to Brighton.

It frequents the paths and borders of woods, and also, it is said, heathy places.

The perfect insect appears the end of June or beginning of July. (Mr. Dale once, namely in 1824, took the larva of this species in the New Forest, Hampshire, on the 1st. of June.)

The wings expand to the width of about two inches and a half. The fore wings are of a rich fulvous ground-colour, the base greenish, and the remainder thickly mottled over with black marks, many of which, especially a row, forming almost a continuous waved streak a little within the margin, are more or less of a crescent form. The extreme edge is pale fulvous, and within this are two black lines intersecting a row of black dots, both again intersected horizontally by the black veins of the wings. The hind wings are marked very much in a similar way: the outside edge is slightly concave.

Underneath, the fore wings shew through most of the marks from above, those however at and near the outside edge being much fainter and more indistinct: the ground-colour too is rather paler. The black marks near the tips are faded into rich brown, sometimes spotted with silver. The hind wings are most beautifully variegated with buff, rich ferruginous, and brown, the upper edge near the base being silver, outside which are from five to seven large silver spots, these again succeeded by an interrupted band of nine or ten still larger ones of different sizes and shapes; these by a row of small rusty red spots, some of which have their centres silver, and these again by another row of seven triangular-shaped silver spangles.

The caterpillar, in one of its first stages, is red, which is afterwards exchanged for olive green, with a white line along the back, and white spots on the sides.

The chrysalis is of a reddish colour, with silvery spots. The insect remains in the chrysalis state about a fortnight.

A curious variety of this insect is mentioned by Mr. A. D. Michael, as having been taken near Cromer, Norfolk, in which all the upper side of the fore wings was of a deep brown colour without spots, but with a lighter margin, in which were three or four darker lunules. Other varieties have occurred in which the spots on the wings were more or less confluent.

The plate is from specimens in my own collection.

DARK-GREEN FRITILLARY.

PLATE L.

Argynnis Aglaia,	OCHSENHEIMER. STEPHENS. JERMYN.
" "	DUNCAN. WESTWOOD.
Papilio Aglaia,	LEWIN. DONOVAN. WILKES. HARRIS.
Acidalia Aglaia,	HUBNER.

THE Dark Green Fritillary frequents heaths, woods, meadows, and downs.

Broadway, near Osberton, Nottinghamshire; Buttercrambe Moor, near Stamford Bridge, Yorkshire; near Carlisle, in Cumberland; Sywell Wood, near Northampton, Barnwell, and Ashton Wold, and the neighbourhood of Polebrook, Northamptonshire; Ambleside, in Westmoreland; Crab Wood, near Winchester, Hampshire; the Downs near Arundel, and Brighton, in Sussex; various parts of Suffolk; Hainault Forest, and other places in Essex; near Axminster, Devonshire; and Great Bedwyn and Sarum, Wiltshire. The woods on the banks of the River Dart, in Devonshire, are also localities for this handsome insect, but it occurs not uncommonly in a great many other counties throughout England, from Hampshire to Scotland, in Rosshire, even on the highest hills, near Glen Sheel, and in the Isle of Arran, "both inclusive." In Wales, near Llandudno.

The perfect insect appears in June and July, and continues till August.

The caterpillar feeds on the dog Violet, *(Viola canina.)*

This species expands like the last to the width of from two inches and a quarter to a little over. The ground-colour is of a rather pale but fine fulvous, marked over with numerous black billets, a row of round spots, another of crescents, followed by the margin, two narrow lines running through a row of black spots, both intersected diagonally by the black veins of the wings; the base is dusky, the outside edge nearly straight; the hinder wings are marked in a very similar manner, the whole of the base dusky.

Underneath, the fore wings are marked as above, but the dark marks near the outside edge are much paler; the tips of the wings are paler than the rest, as are the front edges, and there are a few minute faint silver dots near the corner, and one very obscure one inside them. The hind wings dull green over all their inner and larger portion, excepting towards the upper part of the outer side, where the dull fulvous, which succeeds the green in a white band, runs into it. The border is pale greenish yellow; inside this is a row of seven angular-shaped spots, edged on their inner side, crescent-wise, with dull green. These are followed by an irregular row of seven silver spots on the outside edge of the green colour of different sizes and shapes, the centre one very small, and these by about seven other spots and dots of silver, also of different sizes and shapes, over the base of the green.

The female expands to the width of two inches and three quarters, or a little over; the ground-colour is more dull; the base much more extensively and more deeply darker coloured; the dark billets are larger, and those that are open in the male are filled up with black. The hind wings are also much darker at the base.

Underneath, most of the marks are larger than in the male.

The caterpillar is of a blackish colour, with a whitish line down the back, and another on the side, over which is a row of eight small spots.

One variety of this insect has been described as a distinct species, under the name of 'Papilio Charlotta,' (Haworth,) and 'Argynnis Caroletta,' (Miss Jermyn.) It has "the two costal spots on both sides of the fore wings united, and only nineteen instead of twenty-one silvery spots on the under side of the hind wings, several of the ordinary spots at the base being confluent." Dr. Abbot took three specimens of this variety, nearly all alike, near Bedford; and Mr. Dale has another, taken near Peterborough, which on the under side represents on one wing the character of 'Caroletta,' and on the other that of 'Aglaia,' thereby proving it to be, "sans doute," a variety only.

Another splendid variety, of which specimens have been taken near Ipswich and Birmingham, has "the upper surface of the fore wings almost entirely of a dark brownish black, except a bright linear fulvous mark, and beyond it a much smaller mark of the same colour, with a row of faint tawny spots running parallel with the hinder margin. The hinder wings have the markings considerably more distinct. Beneath, the ground-colour of the fore wings is dark ferruginous, and that of the hind wings pea-green, with twenty-one silvery spots."

My friend, John Curtis, Esq., records a variety intermediate between this and the preceding one. Also in the "Magazine of Natural History," No. 26, a pale buff-coloured variety is mentioned with the spots and markings very faint.

The engraving is from specimens in my own collection.

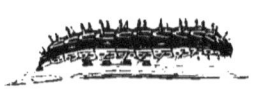 

QUEEN OF SPAIN FRITILLARY.

PLATE LI.

Argynnis Lathonia,	FABRICIUS. OCHSENHEIMER
" "	LEACH. STEPHENS. CURTIS.
" *Latonia*,	ZETTERSTEDT.
Papilio Lathonia,	LINNÆUS. LEWIN. DONOVAN.
" *Principissa*,	LINNÆUS.
" *Lathona*,	HÜBNER.
Issoria Lathonia,	HÜBNER.

IN June, in the year 1803, Dr. Abbot has recorded that he took this rare insect; and the late Mr. J. F. Stephens captured one on the 14th. of August, in the same year. Two specimens were taken, and others seen, at Harleston, near Norwich, in 1846, and three near Dover, the same year. Two on the race-course near Ipswich, Suffolk, in 1851; and two or three pairs at Jagger, near Colchester, in the same year, as R. B. Postans, Esq. has informed me. Stoke-by-Nayland, in Essex, it is also given as one of its localities, as are likewise the neighbourhood of Wisbeach, and near Gamlingay, Cambridgeshire; Halvergate, Norfolk: Chesham, in Buckinghamshire; Brighton, in Sussex; Birch Wood, Ramsgate, and Dover, Kent; Battersea Fields, near London; and Hertford. In Ireland, one at Killarney.

When the summer has fairly set in, with all its gay delights for those who can appreciate them,

"Et nunc omnis ager, nunc omnis parturit arbos,"

when not only botanical, but entomological treasures are abundantly brought forth, then is the time for the appearance of the Queen of Spain Fritillary. The middle of August seems to be its proper time. It has been taken in September. It is believed to be double-brooded, and some individuals of the latter of the two are said to live through the winter.

The perfect insect appears in August and September.

The caterpillar feeds on the heart's-case. *(Viola tricolor,)* the saintfoin, *(Onobrychis sativa,)* and the burrage, *(Borago officinalis.)*

The expanse of the wings is about two inches; their colour is fulvous, with many distinct spots, most of them of a round shape, those at the tip uniting with the dark margin, and enclosing several small paler buff patches; the base dusky. The hind wings are of the same general character, their base also dusky.

Underneath, the fore wings have nearly the same markings as those on the upper surface, but at the tip is a broad ferruginous patch, at the base of which is a silvery spot, succeeded by two small eyes, between which and the margin are several oval-shaped silver patches. The hind wings are buff, varied with reddish brown, with numerous silver patches of different sizes and shapes, of which there are about fourteen between the base of the wings, and a row of seven dark brown eyes with silvery pupils, between each of which and the margin of the wing is a large silvery patch.

The caterpillar is said to be greyish brown, with a white line spotted with black along the back, and two yellowish brown lines along the sides; the spines and legs pale yellow.

The chrysalis is varied with brown and dull green, interspersed with metallic spots.

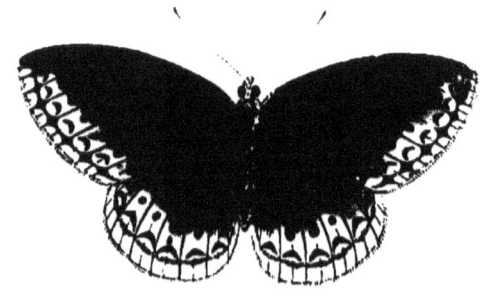

VENUS FRITILLARY.

PLATE LII.

Argynnis Aphrodite,	BREE. WESTWOOD.
Papilio Aphrodite,	FABRICIUS.

THIS is an American species, but it is unquestionable that a specimen was taken in an undoubted wild state, so to speak, in Upton Wood, a few miles from Leamington, Warwickshire, by James Walhouse, Esq., of that place. How it came from the "Far West" is now an undiscoverable mystery. This grand capture occurred in the summer of 1833.

The expanse of the wings is nearly three inches and a quarter. The fore wings are of a rich fulvous colour, spotted and chequered over with black. The hind wings are of the same general ground colour, with very similar markings.

Underneath, the ground-colour is buff, tinged with pink, the tips greenish, the dark marks shewing through. The hind wings are bronze green, but dark at their base, and lighter towards the outside; a row of semicircular silver spots follows the margin, and there are numerous other silver spots.

The engraving is from the figure in Mr. Westwood and Mr. Humphreys' work.

SILVER-WASHED FRITILLARY.

PLATE LIII.

Argynnis Paphia,	FABRICIUS. OCHSENHEIMER. STEPHENS.
" "	WESTWOOD. CURTIS. DUNCAN.
Papilio Paphia,	LINNÆUS. LEWIN. DONOVAN.
" "	WILKES. HARRIS.
Argyronome Paphia,	HUBNER.

THIS is a plentiful species in woods, in the south of England especially, but it extends northwards also to Scotland. A few of its localities are the "Dukeries," near Osberton, Nottinghamshire; Hainault Forest, Essex; Barnwell and Ashton Wold, and the neighbourhood of Polebrook, Northamptonshire; Looe, in Cornwall; near Falmouth it is rare; in Langham Woods, near Stoke Nayland, Suffolk, very abundantly; also, rather uncommonly, near Great Bedwyn and Sarum, Wiltshire, as J. W. Lukis, Esq. tells me; Brighton, in Sussex; the Isle of Wight, Bisterne, and Lyndhurst, in Hampshire; and Lyme Regis, Dorsetshire, where I have also taken it. It is abundant in the woods of Essex, and at Broadway and near Milstead, Kent. It occurs likewise in Sywell Wood, near Northampton, as the Rev. D. T. Knight has informed me, and in various other localities throughout the kingdom, extending in its range even into Scotland. In Yorkshire, at Sutton-on-Derwent, Buttercrambe Moor, Allerthorpe, Stockton, Sand Hutton, etc. In Westmoreland, at Ambleside; and in the woods on the banks of the River Dart, in Devonshire, as James Dalton, Esq., of Worcester College, Oxford, has written me word.

The perfect insect appears the beginning of July.

The caterpillar feeds on the violet, (*Viola canina,*) nettle, and raspberry.

This Fritillary expands in the width of its wings from about two inches and three quarters to nearly three inches. The fore wings are rich fulvous, with numerous blackish spots and bars, the latter horizontal, and of the former there are three rows, following the outside edge of

the wing, the inner row the largest sized, and the outer one on the edge with a dark line running through them. The hind wings are of the like ground-colour, with also three rows of larger spots, the inner rounded, the next bell-shaped, and within the three two waved blackish lines, meeting together near the lower part of the inner one.

Underneath, the fore wings are paler in colour, the outer corner dashed with metallic green; of the dark marks, some shew through, and of others only the outlines; those towards the outer edge of the wing are nearly obliterated. The hind wings are pale metallic green, with two short waves of silver near the base, a third tinged with purple running across the wing, and another still more tinged with purple follows the margin; between these two last is a row of darker green spots, with lighter centres, and another of green half-moons, the latter outermost.

The female is larger than the male, without streaks, the spots larger, the fulvous colour less bright, and tinged with green.

The caterpillar is light brown, with a row of yellow spots on the back; it is covered with long spines, the two next the head being longer than the rest.

The chrysalis is described as being grey, with gilt tubercles.

One variety, a female, taken by J. C. Dale, Esq., has the upper surface of the wings very dark, with some whitish spots at the tips of the fore wings.

In one, figured by Hubner, the wings on the right hand side are of this variety, and those on the left as in the ordinary examples.

Mr. Westwood adds, "A still more remarkable specimen has been figured by M. Wesmaël, in the fourth volume of the Bulletin of the Academy of Brussels, in which the right wings were those of the male type, except that the marginal row of spots were as large as in the female; the left fore wing exhibited a complete 'melange' of the male and female, as well as of the variety and typical individuals, the ground-colour being fulvous, as in the male, but the markings, especially at the tip, dark as in the female, with the white spots of the variety; upper side of the hind wings entirely coloured as in the variety."

Another specimen of this kind is mentioned by Ochsenheimer, the right wings of which were those of a male, and the left those of a female.

In Loudon's "Magazine of Natural History" the occurrence of a similar one in England is noticed.

The engraving is from specimens in my own cabinet.

LARGE COPPER.

PLATE LIV.

Lycaena dispar,	CURTIS. DUNCAN. WOOD. SWAINSON.
Papilio dispar,	HAWORTH. KIRBY AND SPENCE
" *Hippothoë,*	LEWIN. DONOVAN.
" " *var.*	ESPER.
Polyommatus dispar,	BOISDUVAL.
Chrysophanus dispar,	WESTWOOD.

LONG the tenant of the watery wastes which formed the fen districts of Cambridgeshire and the adjoining counties, this fine insect has at last disappeared from what was for ages its secure fastness and its safe stronghold. The industry of man has stopped the "Meeting of the waters," or rather the very waters themselves have been employed against themselves, and condensed in the steam-engine, have driven all before them, as if with the force of the rising tide of the ocean. Science, with one of her many triumphs, has here truly achieved a mighty and a valuable victory, and the land that was once productive only of fever and of ague, now scarce yields to any in broad England in the weight of its golden harvest. Time was, and even abundantly within our own recollection, when it might have been considered a beneficial improvement to introduce a stream of water where none before existed, or to deepen it where it did into a navigable canal, and the engineer who successfully completed the work might well say with a laudable satisfaction—

"Impellitque rates ubi duxit aratra colonus;"

but now the converse is the just subject of boast, and even over the loose surface of the most treacherous morass, the iron way conveys with speed and safety, and to any extent, the mercantile, the physical, and the intellectual wealth of the country. The entomologist is the only person who has cause to lament the change, and he, loyal and patriotic subject as he is, must not repine at even the disappearance of the

Large Copper Butterfly, in the face of such vast and magnificent advantages. Still he may be pardoned for casting "one longing lingering look behind," and I cannot but with some regret recall, at all events, the time when almost any number of this dazzling fly was easily procurable, either "by purchase" or "by exchange," for our cabinets. A goodly "rank and file," from some individuals of which the figures in the plate are taken, I now consider myself fortunate in possessing, for the existing number of indigenous specimens is no more again to be added to by fresh recruits: "Fuit Ilium et ingens gloria"—

"The light of other days has faded, and all its glories past."

Nay, further, not only is it, or rather was it, for it is now, as I have said, extinct, extremely local, but it has always hitherto been believed, like the Grouse, to be peculiar to Britain, being not found elsewhere. These are inexplicable facts in Natural History, into the consideration however of which the limits of my space prevent me from entering. Mr. H. N. Humphreys however states that he took a specimen, which appeared to be identical with it, in the Pontine marshes between Rome and Naples.

The "Fen Districts" of Cambridgeshire and Huntingdonshire, and other congenial places in Norfolk and Suffolk, such as Holme Fen, Whittlesea Mere—now no longer a Mere, Bardolph Fen, and Benacre, were the localities of this fine fly. It was quick and active on the wing, flying among and about the reeds.

It appears, that is to say, used to appear, at the end of July and the beginning of August.

The food of the caterpillar was the water-dock, (*Rumex palustris*.)

This species measures in the expanse of its wings from a little under to a little over an inch and a half. The fore wings are of a splendid copper-colour, with a black edging to the outside of the wing, widest at the upper corner, from whence it decreases; there is a black oblong spot in the centre of the wing, near the front edge; some of the spots from the under side shew faintly through in some lights. The hind wings are of the like colour and markings, only that the outside black border is indented and equally wide all along, except at the uppermost part, where it is narrower; the whole of the inner edge, from the base downwards, is dusky black.

Underneath, the fore wings are pale orange, the outside edge blue ash grey; there are two large and one small black spots, placed horizontally, in a row from near the base, their edges bordered with a line of still paler orange; the inner one is the smallest: these are succeeded by a transverse row of seven others, of smaller comparative sizes, three and four, and there is a row of small faint black crescents on the inner edge of the grey band. The hind wings are silvery greyish blue, brightest near the base, followed by a broad oblong orange bar, which again is edged by the grey and a row on each side of black dots, the inner ones larger than the outer; on the grey part the black oblong mark shews through, there being within it five or six black dots, one near the upper edge large, and an irregular row of about nine others near the orange band.

In the female the copper on the fore wings is of a deeper colour, with a wide dusky black border running over part of the front edge: the lower part of these wings near the base is also dusky black, and there are two large black spots placed horizontally near the front edge, and a transverse row of six or seven other ones towards the margin, the middle ones being large and elongated. In the hind wings the copper is almost entirely hid by blackish brown, excepting a broad bar of the former near the outer edge, indented by the black on its outer margin, and running up into the black, which it intersects in narrow streaks: the outer edge is bluish white.

The caterpillar is described by the late Mr. J. F. Stephens as of a bright green colour, with innumerable white dots.

The chrysalis was "at first green, then pale ash-coloured, with a dark dorsal line, and two abbreviated white ones on each side, and lastly sometimes deep brown."

SMALL COPPER.

COMMON COPPER. COMMON SMALL COPPER.

PLATE LV.

Lycaena Phlaeas,	FABRICIUS. OCHSENHEIMER. LEACH.
" "	STEPHENS. CURTIS. DUNCAN. WOOD.
Papilio Phlaeas,	LINNÆUS. HAWORTH. LEWIN.
" "	DONOVAN. HARRIS.
Chrysophanus Phlaeas,	HUBNER. WESTWOOD.
Polyommatus Phlaeas,	BOISDUVAL.

This butterfly occurs throughout Europe, and in Asia, and also, or a closely allied species, in America.

It is a common insect with us, and generally distributed throughout the country. It is an exceedingly elegant object on the wing, as it flits from flower to flower, its showy colour, though it is so small, attracting the eye. It seems to be fond of attacking and fighting with any of its fellows that approaches, but the difference may be more apparent than real—a mere "passage of arms" essayed in the exuberance of the happiness of the ephemeral little creature.

There are two or three broods in the year, and they appear early in April and May, in June, and in August.

The caterpillar feeds on the sorrel, (*Oxalis acetosella.*)

The fore wings, which expand from a little over an inch to a little over one and a quarter, are of a resplendent copper colour, with from eight to ten black spots of different sizes and shapes on their central part; of these, the two or three nearest to the base of the wing are placed transversely. The front edge of the wing is narrowly margined with brown, and the outer edge broadly so; the fringe is buff. The hind wings are dark blackish brown, with a copper bar at the lower side, edged on its lower part with black crescent-shaped spots; the fringe is buff.

SMALL COPPER.

Underneath, the fore wings are of an orange colour, the black spots shewing through, and slightly edged with pale buff. The outer margin, which is dull buff, is edged on its inside with several faint dark-coloured crescents near the lower corner, fragments of the broad border on the upper side. The hind wings are also a sort of buff, with numerous minute and nearly obsolete golden brown specks placed in irregular rows, and with an obscure narrow bright orange band parallel with the hind margin. There are short tails to the wings, which are indented within them. The body is black on the upper side, with some tawny down about the head and thorax, and greyish buff-coloured beneath.

The female resembles the male.

The caterpillar is green, with a pale line along the back, and also one on the side.

Specimens differ considerably in size, and also in the depth of their colour:—

One has been taken, in which the copper band on the hind wings is wholly obliterated.

Another had the copper colour on all the wings exchanged for milk white.

A third had the black parts replaced by milk white.

The plate is from specimens in my own cabinet.

BRIGHTON ARGUS.

PLATE LVI.

Lycæna Bætica,   AUCTORUM.

NOTHING is more curious in the whole range of natural history than the way in which certain insects, as well as certain birds and animals, exclusively belong to particular countries, or to particular districts of such countries. Some are peculiar to one continent, some to another; some to this island, some to that; few indeed, if any, are of universal distribution, and though some are met with in divers quarters of the earth, yet by far the greater majority 'have their "bounds which they do not pass," or if they do so on some occasion or in particular instances, the exception only proves the rule, which is all the other way.

The insect before us is an instance of this. It is seen on the opposite coast of France, and in various parts of the continent of Europe, as likewise in the Channel Islands, on which account also it has a claim to a place on our list of native insects.

It has been taken twice, and as yet twice only, in this country; on the former of the two occasions by Mr. Mc'Arthur near the Chalk Downs, in the neighbourhood of Brighton, on the 5th. of August.

That is the month for the appearance of the perfect insect.

The fore wings are of a dull brownish black colour, with a patch of iridescent purple on the middle part.

The hind wings are of a dull fulvous brown colour, with a tint of dull blue on the inner side, and three black spots on the lower corner, the centre one larger than the others, and very distinct, with a white vein round it; the others more dull in each of these respects. There are short tails to these wings after the fashion of the Hairstreaks.

Underneath, the wings are dove-coloured brown, elegantly mottled and streaked with a paler shade of the same, with a bar of the latter within the outer margin, and a fulvous red mark at its outside corner, in which are two bright rich dark emerald spots.

The tails to the wings, as well as the purple colour in the fore wings, shew this butterfly to be closely allied to the family of the Hairstreaks, so that it can scarcely be called a "True Blue."

For the specimen of this insect from which the figure on the plate is taken, I have to thank Mr. Cooke, the naturalist, of Oxford Street, London; and my obliging friend, Frederick Smith, Esq., of the British Museum, by whose kind intervention it was obtained for me.

MAZARINE BLUE.

PLATE LVII.

Polyommatus Acis,	STEPHENS. CURTIS. WOOD.
" "	DUNCAN. WESTWOOD.
Papilio Acis,	ERNST.
Lycaena Acis,	OCHSENHEIMER.
Nomiades Acis,	HUBNER.
Papilio Argiolus,	ESPER. HUBNER.
" *semiargus,*	BORKHAUSEN.
" *Cymon,*	LEWIN. HAWORTH. JERMYN.
Lycaena Cymon,	LEACH. SAMOUELLE.

THIS, in conjunction with a former species, gives us an entomological "Acis and Galatea," who

> "Merry, harmless, free, and gay,
> Dance and sport the hours away;
> For them the zephr blows, for them distils the dew,
> For them unfolds the rose, and flow'rs display their hue."
>
> "Where shall I seek the charming fair?
> Direct the way kind genius of the mountains;—
> Seeks she the groves?"

It is rather the "pleasure of the plains" that will reward your search, and even there you must wander far and wide, for it is but at distant intervals both of time and space that our present butterfly is to be met with.

This very interesting and valuable insect used formerly to be taken in tolerable plenty by J. C. Dale, Esq., in his parish of Glanville's Wootton, Dorsetshire, but now it is never seen there. It occurs also near Sarum, Wiltshire, and in Sywell Wood, near Northampton, as the Rev. D. T. Knight, of Earl's Barton, informs me. Other localities given are Lincolnshire, as at Epworth near Bawtry, and Norfolk, Cherry Hinton, and various parts of Cambridgeshire, Hampshire, and Worcestershire; also near Hinkley, Leicestershire; and one was once taken in Coleshill Park, Warwickshire, by the Rev. W. T. Bree.

It is out the latter end of July.

The fore wings, near the centre of which is a dark spot, are of a dark purple blue colour, the front edge thinly edged with white; the outer margin is narrow, and dark brown, which colour runs up into the wing along the veins; the fringe is white. The hind wings are also fringed with white, and their outer margin is also narrow, and dark brown.

Underneath, the fore wings are of a uniform bronzed grey colour, the base bluish grey; the spot shows through, edged with white, and beyond it is a waved row of seven spots ringed with white, the middle ones larger, the end ones smaller and much less distinct, especially on the lower side, where two are adjoined; these spots vary in number in different individuals; in some there are also one or two eyed spots near the base.

The female differs from the male in having the upper side of all the wings of a very dark copper brown, the basal part alone purpled.

The female is figured from a specimen in my own collection.

LARGE BLUE.

PLATE LVIII.

Polyommatus Arion.	LATREILLE. STEPHENS. CURTIS.
" "	WOOD. DUNCAN. WESTWOOD.
Papilio Arion,	LINNÆUS. HAWORTH. LEWIN.
" "	DONOVAN. HUBNER.
Lycæna Arion,	OCHSENHEIMER. LEACH.
Nomiades Arion.	HUBNER.

LOCALITIES for this insect are Charmouth, Dorset, where my brother, Beverley R. Morris, Esq. once took one, which I saw immediately afterwards; the Cliffs near Dover, Kent; the Downs near Marlborough, Wiltshire; and those near Glastonbury and Langport, Somersetshire, where Mr. Quekett has taken it in some profusion; also near Cheltenham; the Mouse's Pasture, near Bedford, where Dr. Abbot first took it, and Mr. Dale again in 1819; Broomham, in Bedfordshire; near Winchester, in Hampshire; Brington, in Rutlandshire; Chatteris, in Cambridgeshire; Brighton, in Sussex; also near Gloucester; and Wigsworth and Barnwell Wold, near Oundle, Northamptonshire, where, in company with my hospitable host, the Rev. William Bree, Curate of Polebrook in that neighbourhood, who had in previous years discovered and taken it there in tolerable plenty, I took the considerable number of eleven specimens on the 19th. and 20th. of July, 1852. He only once took one in the neighbourhood of Ashton Wold, a few miles distant, though there are several other fields of exactly a similar appearance to those in which it occurs at the former place. Bolt Head in Devonshire, is another locality for it.

It appears in June and July.

The Large Blue expands in width from a little over an inch and a half to nearly one and three quarters. The fore wings are of a deep blue colour, with a large cluster of oblong black spots near their centre; the outer edges are dark coloured; the fringe of the wings is white. The hind wings are of the like deep blue colour, bordered with blackish brown.

Underneath, the colour is dark grey; the black oblong spots are replaced by rounded ones, with white orbits, and two rows of smaller dots of a similar character follow the black line that bounds the outside of the wing, which is succeeded by a white fringe: the base of these wings is blue. The hind wings are much spotted, and their base is blue considerably diffused; two irregular rows of black spots, with white edges, run across the middle part, and there is one single one near the base; a double row of black angular-shaped spots, based with whitish, follows the outside marginal line: the fringe whitish.

The female is of a duller colour than the male, the spots larger and more numerous, and the dark margin of the wings broader.

This insect varies very much in the number and size of the spots. Some specimens are almost wholly immaculate, and others are gradually spotted 'seriatim' more and more till the wings are strikingly marked.

The figures are from specimens in my own cabinet.

HOLLY BLUE.

AZURE BLUE.

PLATE LIX.

Polyommatus Argiolus,	LATREILLE. STEPHENS. CURTIS.
" "	WOOD. DUNCAN. WESTWOOD.
Papilio Argiolus,	LINNÆUS. HAWORTH. DONOVAN.
" "	LEWIN.
" *Acis,*	HUBNER.
" *Cleobis,*	ESPER.
" *Argus marginatus,*	DE GEER.
Agriades Argiolus,	HUBNER.
Lycaena Argiolus,	OCHSENHEIMER. LEACH.
" "	SAMOUELLE.

This plain but neat species is to be found, as its name imports, in places where the holly abounds. Unlike the other Blues, it flies near the tops of these trees, hovering about them and settling on them very much after the manner of the Hairstreaks.

In Yorkshire it is not uncommon. I have known it in former years in very considerable plenty at Fairfield, near Broomsgrove, Worcestershire; other localities for it are Castle Eden Dene, Durham; Dartford, Kent; Ripley, Surrey; Epping Forest, Essex; Newcastle, Northumberland; Hammersmith, near London; Allesley, near Birmingham, and other places, in Warwickshire; the Isle of Wight, and in Hampshire; and also in various places in Norfolk, Suffolk, and Devonshire. Brighton and Lewes, in Sussex; Blean Wood, near Canterbury; Looe, in Cornwall; Weston-super-Mare, in Somersetshire. In Wales, near Llandudno.

It appears to be double-brooded, being taken so early as the middle of April, as also in May, June, July, and the end of August—the charming summertide, when you cannot but, with the butterflies you seek, "love the merry merry sunshine," and wish that it were always summer.

The caterpillar feeds on the buck-thorn. (*Rhamnus catharticus,*) and the holly, (*Ilex Aquifolium.*)

In this species the wings expand to nearly an inch and a quarter in width. The fore wings are of a uniform rather dull pale blue; the hind wings the same: the fringe of all the wings is white.

The female differs from the male in being generally of a smaller size, the blue of a paler hue, and the fore wings broadly margined with black or blackish brown. The hind wings are also marked with a row of black or dark brown spots within the margin, which are sometimes so large as to be almost confluent. Their upper edge is broadly margined also with black.

Underneath, the whole surface is of a uniform pale silvery greyish blue, crossed towards the outside of the fore wings with an irregular row of small black dots. The hind wings have also a few minute dots on different parts of their surface, the principal one on the middle of the upper edge: the spots as well as the dusky markings vary considerably in size, and the former in number in different individuals.

The caterpillar is of a greenish yellow colour, with a bright green line along the back, the head and legs being also black.

The chrysalis is smooth, brown and green, with a dark line along the back.

The figures are from specimens in my own cabinet.

LITTLE BLUE.

BEDFORD BLUE.

PLATE LX.

Polyommatus Alsus,	Stephens. Curtis. Duncan.
" "	Wood. Westwood.
Papilio Alsus,	Gmelin. Lewin. Donovan.
" *minimus.*	Esper. Schaeffer. Villars.
" *pseudolus,*	Borkhausen.
Nomiades Alsus,	Hubner.
Hesperia Alsus,	Fabricius.

I HAVE taken this diminutive butterfly in tolerable plenty at Pinney Cliff, Devonshire; near Lyme Regis, and also near Charmouth, Dorsetshire; likewise near Sittingbourne, in Kent. In Yorkshire, it has been met with at Wadsworth and Brodsworth, near Doncaster, Langton Wold, Londesborough, by myself, and in other parts. Other localities for it are the neighbourhood of Newmarket and Cherry Hinton, Cambridgeshire; near Great Bedwyn and Sarum, Wiltshire, in isolated places near woods, as J. W. Lukis, Esq. has informed me, as likewise near Amesbury; and Hainault Forest, Essex. Also South Creek, Norfolk; Brandon Warren, Suffolk; Brighton, in Sussex; Weston-super-Mare, in Somersetshire; Ranmore Common, near Dorking, Surrey; Dover, Birch Wood, and Darenth Wood, Kent; near Andover, Winchester, and the Isle of Wight, in Hampshire; Darlington, in Durham; between Woodstock and Enstone, Oxfordshire; near Cheltenham, in Gloucestershire; and near Hertford. In most of the northern counties of Scotland, and at Ardrahan, in the county of Galway, in Ireland, as A. G. More, Esq., has written me word. In Wales, in Gloddaeth Wood near Llandudno.

The perfect insect appears at the end of May or beginning of June, and keeps out for a considerable time. It is often seen on the sides of disused chalk-pits, and on grass-grown cliffs, where, a veritable

"Gay being," in revels in the concentrated sunshine which there glows with an unsubdued heat such as the race of butterflies rejoices in.

The caterpillar feeds on the milk-vetch, (*Astragalus cicer.*)

The expanse of the wings is between three quarters of an inch and an inch. The upper surface of the fore wings is obscure dark brown, more or less glossed with blue, chiefly at the base. The hind wings are likewise of a dull dark brown colour. The fringe of the wings is white.

Underneath, the fore wings are of a pale silvery ash-colour, with a small black dot near the front edge, and between this and the hind margin is a transverse row of small black spots with white rims, the two lower being more confluent. The hind wings, which are of the like ground-colour, have three or four of the eyed spots irregularly placed on their inner part, beyond the middle of which is a waved row of seven or eight similar spots, and on the margin is a black spot near the lower corner.

The female is duller in colour than the male.

The caterpillar is green, with yellow lines on the sides and the back.

The figures are from one in my own cabinet, and from others in that of Mr. Allis, of Osbaldwick, near York.

SILVER-STUDDED BLUE.

PLATE LXI.

Polyommatus Argus,		STEPHENS. DUNCAN.
"	"	WESTWOOD. WOOD.
"	*Alcippe,*	KIRBY, MS., (*var.*)
"	*maritimus,*	HAWORTH, MS., (*var.*)
Papilio Argus,		LINNÆUS. LEWIN. HAWORTH.
"	*Idas,*	LINNÆUS, (*female.*)
"	*Argyrognomon,*	BORKHAUSEN, (*var?*)
"	*Leodorus,*	ESPER, (*var.*)
"	*Argiades,*	ESPER, (*var.*)
Hesperia Argus,		FABRICIUS.
"	*Acreon,*	FABRICIUS, (*var?*)
Lycæna Argus,		LEACH. OCHSENHEIMER. HUBNER.
Lycæides Argus.		HUBNER.

THIS fly is not uncommon near Sarum, Wiltshire; also in Sywell Wood, near Northampton. I have taken it in Devonshire, at Pinhay Cliff, near Lyme Regis. It is also found in Coombe Wood and Darenth Wood, Kent; and Wells, in Somersetshire; at Wood Hay Common and Bisterne, Hampshire; on Coleshill Heath, Warwickshire; Parley Heath, Dorsetshire; Langwith, near York; Brighton, in Sussex; Looe, in Cornwall; Holt, in Norfolk; and Ripley Green, Surrey; as likewise in various other localities in the south of England.

It is a very pretty and interesting, though rather plain species; and pleasant it is to watch it as it wanders about to "bid good-morrow to the flowers," in the height of summer, when you are glad to lie down on some grassy bank and gaze upon the plants or the insects which surround you, listening the while to the murmur of the tinkling rill, or it may be the gentle rippling of the tide over the pebbled beach, every sight and every sound full of present and inexpressible enjoyment, and recalling perhaps also other times and other scenes and passages connected with them, which, alas! cannot be otherwise recalled than by the memory, too retentive, and yet not retentive enough.

The perfect insect appears about the middle of July.

The caterpillar feeds on the broom, (*Sarothamnus scoparius,*) saintfoin, (*Onobrychis sativa,*) and other species of the clover, (*Trifolium,*) and allied kinds.

In this species, which measures from an inch to an inch and a quarter across, the fore wings are of a silvery blue colour, the front part verging to white; there is a minute speck near the centre, and the edge is black; the fringe white, with sometimes a row of orange dots at the edge. The hind wings are of a similar colour, but the dark edge is wider, seemingly a tissue of spots, and the fore part has also a dark margin; the body above is clothed with silvery and blue down. "The antennæ are black, with white rings, the upper side of the club black, and the lower fine orange."

Underneath, the fore wings are of a pale greyish lilac tint, the base saturated with blue; the spot shews through larger, and beyond it is a waved row of black spots, with white rims, the ground-colour under them being paler than the rest of the wings, followed by two rows of minute and faint ones on a pale orange ground, followed by a black line, and this by the white fringe. The hind wings are marked in much the same way, but there are two additional small eyes near the base and one at the front edge; the pale ground under the row of spots is nearly white, followed by a band of clear orange, with a row of dots on its inner and outer edges, the latter larger and more distinct than the former.

The female, which is larger than the male, has the fore wings of a copper brown colour, sometimes faintly tinted with blue, with a line of obscure orange spots near the margin; the fringe reddish buff, white at the tip and on the front edge. The hind wings are marked in a similar manner, but the row of orange spots is larger and more distinct, and farther within the margin; in some specimens these wings especially have a faint suffusion of blue.

Underneath, the fore wings are dark ash grey, with a central eyed spot, then a waved row of the like, then a partial white streak, then a row of black spots, shaded off on the edges, followed successively by an orange band, a row of small black dots, an interrupted white line, a narrow black line, and a greyish white fringe. The hind wings have much the same markings, but there are several spots over their inner portion, and there is a row of bright metallic green streaks along the outside of the orange band.

The caterpillar is described as being "of a dull green colour, with whitish tubercles, and a blackish head and legs, a line down the back and sides, and oblique marks on the latter of a dark red colour bordered with white."

The chrysalis is said to at first green, and afterwards brown.

A variety of this species, described as a separate one under the name of 'Polyommatus Alcippe,' has "the wings narrower, blue above, with a broad black margin to all the wings, the under side of the male of a deep greyish or drab colour; the ocelli very distinct in the female, and the oblique series on the posterior wing consisting of four."

In another taken by the late Mr. Haworth, and named by him 'Polyommatus maritimus,' "the ocelli on the disc of the under side of the wings are elongated into those on the middle of the wing, being almost confluent with the following row of spots."

Another, taken by the late Mr. Hatchett, had the upper surface of all the wings of a pale fulvous tawny-colour.

The engraving is from specimens in my own cabinet.

COMMON BLUE.

PLATE LXII.

Polyommatus Alexis,	LATREILLE. STEPHENS. CURTIS.
" "	WOOD. DUNCAN. WESTWOOD.
" *Labienus,*	JERMYN, (*var.*)
" *Thestylis,*	JERMYN, (*var.*)
" *Lacon,*	JERMYN, (*var.*)
" *dubius,*	KIRBY, MSS., (*var.*)
Papilio Alexis,	HUBNER. WIENER.
" *Argus,*	WILKES. DONOVAN. HARRIS.
" *Hyacinthus,*	LEWIN. HAWORTH, (*var.*)
" *Icarus,*	VILLARS. HAWORTH. LEWIN.
Lycæna Dorylas,	LEACH. SAMOUELLE.

This is one of the commonest of our native species, and appears to be distributed throughout the kingdom.

In the "Journal of a Naturalist," it is thus accurately noticed by Mr. Knapp:—"We have few more zealous and pugnacious insects than this little elegant butterfly, noted and admired by all. When fully animated, it will not suffer any of its tribe to cross its path, or approach the flower on which it sits, with impunity; even the large admirable Atalanta at these times it will assail and drive away. Constant warfare is also kept up between it and the Small Copper Butterfly, and wherever these diminutive creatures come near each other, they dart into action, and continue buffeting one another about till one retires from the contest, when the victor returns in triumph to the station he had left. Should the enemy again advance, the combat is renewed; but should a cloud obscure the sun, or a breeze chill the air, their ardour becomes abated, and contention ceases. The pugnacious disposition of the Argus Butterfly soon deprives it of much of its beauty, and unless captured soon after its birth, we find the margins of its wings torn and jagged, the elegant blue rubbed from the wings."

This butterfly is out in June, and there is a second brood in August. The caterpillar is found the end of April, and in July and August.

It is said to feed on the wild strawberry, (*Fragaria cesca*,) the wild liquorice, and different kinds of grasses.

The Common Blue averages in the expanse of its wings from a little over an inch to an inch and a quarter. The fore wings are of a fine lilac blue, margined on the outer edge with a thin black line; the fringe white. A similar description applies to the hind wings.

Underneath, the front wings are of an ash-colour; towards the base is one ocellated spot, and beneath it a black line, then another spot, then a transverse row of six others, and then two rows of smaller and fainter ones; the lower ones of each row with some pale orange marks between them; these are succeeded by a narrow black thin line at the edge of the white fringe. The hind wings are irrorated about the base with silvery blue, and are spotted very much in the same way as the fore wings, but there is a bidentate spot below the centre, and the orange spots outside this are large, continuous, and distinct, following the margin of the wing. The body is clothed with long downy hair, of a bluish white colour.

In the female the blue of the fore wings is almost wholly obscured with blackish brown, which latter colour forms a distinct border at the outer edge, within it being a row of orange spots more or less distinct in different specimens: the fringe is white. The hind wings are similarly marked, except that the black edge is supplanted by a narrow black line, within which is a blue one, with a row of black spots continuous with the orange ones.

Underneath, the markings resemble those in the male, but they are brighter and more distinct.

The caterpillar is of a bright green colour, with a dark line along the back, adjoining which are rows of yellow spots.

This species is much subject to variety, both in the number and size of the eyes on the under surface of the wings, and the markings on the upper, and hence has acquired many an "alias," as shewn by its synonyms and supposed distinctions.

Some individuals exhibit the double appearance of the male and the female, viz., the wings on each side representing each sex.

Others have the sides not correspondingly alike.

Some differ in form from the rest, the tips of the wings in the females being rounded, or acute.

Some females have the upper wings nearly as blue as in the male, with a black spot, while in others they are nearly entirely blackish brown.

One variety, of a very small size, described as a separate species by the name of 'Polyommatus Labienus,' had "the upper side of the wings of a very pale lilac blue, and the spots on the under side very small and pale, the inferior spot at the base of the fore wings obsolete, only five spots in the curved row beyond the middle of the discoidal cell, and the fulvous lunules almost obsolete, the two basal spots on the costa of the hind wings large and black."

Another, a large female, the 'Polyommatus Thestylis' of Jermyn, in which the blue of the upper surface of the wings was more than ordinarily extended, had the front wings with a large blackish spot, obscurely engirdled with white, the hind wings with a similar spot near the margin, and the number of eyes in all the [wings], varying considerably.

Another variety is the 'Polyommatus Lacon,' also of Miss Jermyn, "in which the disc of the wings beneath is only marked with a triangular spot; the hind margin of the anterior with a few indistinct dusky marks, and of the posterior ones with a fulvous band, terminated internally with a series of black wedge-shaped spots, and externally with black dots on a white ground."

Another had "the two spots towards the base of the fore wings on the under side obsolete, and the upper side of the wings of the female more strongly saturated with blue."

Some males have the wings very transparent, and of a more than ordinary silvery hue, and some females "very blue, with very distinct red lunules."

The engraving is from specimens in my own cabinet.

CLIFDEN BLUE.

DARTFORD BLUE.

PLATE LXIII.

Polyommatus Adonis,	STEPHENS. CURTIS. WOOD.
" "	DUNCAN. WESTWOOD.
Papilio Adonis,	OCHSENHEIMER. LEACH.
" "	SAMOUELLE.
" *Ceronus,*	HUBNER.
" *Bellargus,*	ESPER. VILLARS.
" *Argus,*	DONOVAN.
Hesperia Adonis,	FABRICIUS.

THE unfeeling and heartless manner in which Mr. Charles Dickens relates the gratuitous destruction, by the robbers, of poor Grimaldi's "Dartford Blues," in revenge, as it would appear, for the rest of the intended plunder having been timely removed, must for ever lower him in the estimation of every high-souled—entomologist. True indeed it is that he uses language not altogether inappropriate, in treating of the loss—language which, did it express the feelings of his heart, might be accepted as displaying some degree of commiseration for so lamentable a calamity; but the acute perception of the entomologist will at once tell him that the sympathy is but feigned, the pity but a mockery, the pretended commiseration a mere delusion, betokening an utter want of feeling on a subject which ought instinctively to call forth the deepest emotion. 'T is easy to see through the hollow speciousness: 'hic nigri est succus loliginis;' which translated into plain English is—Mr. Dickens, I am quite sure, is no entomologist.

This most lovely insect, whose beauty is imported by its specific name, is abundant in some seasons in the vicinity of Croydon, Surrey, as Mr. C. Miller informs me: it also occurs in some parts of Suffolk, and other southern counties, as Brighton, in Sussex, Weston-super-

Mare, in Somersetshire. It too frequents the districts of the chalk formation.

It appears to be double-brooded, some appearing at the end of May, and others in the middle of August.

The expanse of the wings is from an inch and a quarter to nearly one and a half. The fore wings are of a splendid polished azure blue, the fringe white, intersected by the veins, edged interiorly by an attenuated black line. The hind wings are of a similar appearance.

Underneath, the fore wings are of a dark ash grey, with one eyed spot towards the base, then a large dark one near the centre, then a waved row of six, and then two other faint rows. The hind wings are powdered with silvery blue at the base; near the front margin are three eyed spots, and below these two others; the outer one near the centre with a larger white circumference, and a comet-like tail pointing downwards. Between these and the outer margin are two rows of eyed spots, the inner one irregular, and with a white patch about its middle, the outer one following the black line which meets the margin, and is followed by the white fringe. Between the two rows are some pale orange marks with white crescents on their inner edge.

The female has the fore wings of a dark brown colour, the base sometimes marked with blue. There is a white spot near the centre towards the front edge, with a black speck on it. The fringe is pale buff white, chastely striated on the upper portion with blackish brown, and within it is an obscure row of dark dots, tipped interiorly with faint dull white. The hind wings are similarly marked, but there is more blue on their inner portion, and there is a row within the margin of black dots, some of them set in orange, and within these a row of small angular-shaped blue specks.

The caterpillar is described as being green, with rows of fulvous spots along the back.

This butterfly also varies in the number, size, and situation of the spots on the under side, and in some specimens those on the one side do not even correspond with those on the other.

The figures are from specimens in my own cabinet.

CHALK HILL BLUE.

PLATE LXIV.

Polyommatus Corydon,	LATREILLE. STEPHENS. CURTIS.
" "	DUNCAN. WOOD. JERMYN.
" "	WESTWOOD.
Hesperia Corydon,	FABRICIUS. HUBNER. LEWIN.
" "	DONOVAN. ESPER.
Agriades Corydon.	HUBNER.
Papilio Tiphis.	ESPER. (*female.*)
" *Calathis.*	JERMYN. (*var.*)

I BELIEVE I once took a specimen of this elegant butterfly, which frequents the chalk districts, on the Downs a few miles from Lambourne, Berkshire, near Ashdown Park, the seat of Lord Craven, a very singularly situated mansion, a sort of "Oasis in the desert." It occurs abundantly in Epping Forest, in Essex, and near Cherry Hinton, the Gogmagog Hills, and Newmarket, Cambridgeshire, where I have captured it; and is not uncommon near Sarum, and also at Martin's Hill, near Great Bedwyn, Wiltshire, where J. W. Lukis, Esq. has obtained it, and also near Croydon, Surrey, in some seasons; at Brighton, in Sussex; Blean Down, Weston-super-Mare, and Wells, in Somersetshire; it is plentiful also on the grassy slopes and pastures known as the "Downs" near Hunstanton, Norfolk, as Mr. Robert Marris has informed me; so too at Grange, in Lancashire. Other localities are Dover and Darenth Wood, Kent; Shoreham, Sussex; Prestbury, near Cheltenham, Gloucestershire; Bisterne, near Winchester, and Newport, in the Isle of Wight, Hampshire; different parts of Suffolk; and in Oxfordshire: one was taken near Knowle, Warwickshire; also at Notting Hill, London; and Lulworth Cove, Dorsetshire.

It appears the beginning of July, and is out in August.

The caterpillar is said to feed on the wild thyme, (*Thymus Serpyllum.*)

The Chalk Hill Blue varies in the expanse of its wings from an inch and a quarter to an inch and a half. The male has the fore wings of a most elegant pale metallic blue, with a cast of white; the outer part of the front edge, and a broad border on the outside of the wings, as also the veins, dusky black; there is a row of black dots almost hidden by the wide border; the fringe is white, crossed by the veins, ending in dots, and very wide, except on the front edge. The hind wings are of a similar colour, with a similar broad border on the outer part, the lower being removed, and shewing the black dots distinctly, bordered by a line only of the former on the outside; the dots are more or less ringed with silvery white; the border at the front of these wings is broader than on the outside; the fringe is white, and very wide.

Underneath, the fore wings are pale greyish white, the margin defined by a dusky line, through which the veins run and end in dots; near the middle, towards the base, is a black dot, faintly ringed with white, then another larger oblong central one, then an irregular diagonal row of the like, the middle ones being the largest, and the lower two close together, then another row of smaller and rather fainter ones, succeeded by another: the fringe is white, intersected by the veins. The hind wings are greenish blue about the base, and generally of a pale greyish brown hue, the whitish grey occasionally breaking through, especially near the lower corner; the fringe is white, marked off by a marginal dusky line; these wings are much spotted with a variety of spots and dots, some of them eyed with white, some of the lower ones of a dull orange colour.

The female has the fore wings of a dark bronzed brown, with more or less phosphorescence of blue; near the centre is a small white spot, with a black pupil; the fringe is dull white, the veins crossing it ending more widely: there is a row of obscure light-coloured dots within the margin. The hind wings are of the same ground-colour, but there is more of the blue tinge, at least in some specimens; they also have a small white spot in their centre: the margin is dull white, crossed by the veins, and within it is a row of black dots, partially encircled on the inner side by orange, and these again followed, at least in some individuals, by small bluish triangular-shaped marks.

Underneath, the wings are similarly marked to those of the male, but the ground-colour is very much darker, the spots much larger and more distinct, with white rims, and there are some orange marks within the margin of the fore wings, and a decided row of orange lunules within that of the hind ones.

The caterpillar is green, with yellow lines on the sides and the back.

Many varieties of this species have occurred, the eyes being more or less distinct, and the brown more or less diffused over the wings of the male.

One, described as a species by Miss Jermyn, under the name of 'Polyommatus Calæthys,' has the wings "above brown, with a blue disc and a whitish discoidal spot with a black pupil; beneath, the posterior wings have a discoidal, white, cinctured crescent, with a waved band of seven undulated spots towards the hinder margin."

The engraving is from specimens in my own cabinet.

BROWN ARGUS BLUE.

DURHAM ARGUS. SCOTCH ARGUS.

PLATE LXV.

Polyommatus Agestis,	JERMYN. STEPHENS.
" "	DUNCAN. WOOD. WESTWOOD.
" *Salmacis,*	STEPHENS. WESTWOOD.
" "	WOOD. DUNCAN.
" *Artaxerxes,*	STEPHENS. JERMYN. WOOD.
" "	DUNCAN. WESTWOOD.
Papilio Agestis,	HUBNER.
" *Idas,*	LEWIN. DONOVAN. HAWORTH.
" *Medon,*	JERMYN.
Lycæna Artaxerxes,	LEACH.
" *Idas,*	OCHSENHEIMER. LEACH.
Agriades Agestis,	HUBNER.
Argus Artaxerxes,	BOISDUVAL.
Hesperia Artaxerxes,	FABRICIUS. LEWIN.
" "	HAWORTH. DONOVAN.

THIS insect is common near Great Bedwyn and Sarum, Wiltshire, as J. W. Lukis, Esq. has informed me; and also on the "Downs" near Hunstanton, Norfolk, as Mr. Robert Marris has written me word: I have taken it in tolerable numbers at Pinhay Cliff, Devonshire, near Lyme Regis. It also occurs in various other localities throughout the kingdom; Ramsgate, Dover, and Hythe, Kent; Hastings, Rye, Brighton, Worthing, Little Hampton, and Chichester, Sussex; near Portsmouth, Winchester, Bisterne, and the Isle of Wight, Hampshire; near Birmingham, Warwickshire; Lulworth, Dorsetshire; near Worcester; Shrewsbury, in Shropshire; Manchester, Lancashire; Wells, Blean Down, and Weston-super-Mare, in Somersetshire; Ranmore Common, near Dorking, in Surrey; Castle Eden Dene, Durham, and Seaham Dene near

Sunderland. Flisk, in Fifeshire; near Queensbury and Roslyn Castle; Jardine Hall, Dumfriesshire; King's Park, Salisbury Crag, near Duddingstone Loch, the Pentland Hills, and Arthur's Seat, near Edinburgh, where the 'gaudentes rure camænæ' will present more attractions to the entomologist than the "Modern Athens" itself; also near Perth. In Wales, on the sand-hills near Llandudno.

For the most part this Blue seems to prefer the neighbourhood of the coast.

It is double-brooded, appearing in June and in August.

The caterpillar is found in April and in June.

It is said to feed on different grasses and the wild strawberry.

The expansion of the wings is a little over an inch. The fore wings are glossy brownish black, with a small crescent-shaped black spot near the middle. The margin is narrow and pale whitish grey, with very fine vein lines. The hind wings are of the like ground-colours, with a row of bright orange-coloured crescented spots, largest on the inside part, and nearly obliterated on the outer; in some specimens they are all scarcely discernible.

Underneath, the ground-colour of the fore wings is a chaste grey, with a row, more or less curved, of clear white spots, within which the spots shew through white, sometimes enclosing a black one; outside the white row is another of orange, more or less bright, followed by a slender black line; the inside border of the fringe, which is white. The hind wings are of the same ground-colour, tinted with blue about the base, near which there are three white spots, two others at and near the upper edge, the former the larger, and an irregular row of white spots, followed by another of orange ones, dotted with black on their lower corner, bounded by the black line which makes the inside of the fringe, which is white.

"The caterpillar is green, with a pale angulated row of dorsal spots, and a central brownish line."

The changes in the markings on the wings in this insect, in different latitudes of the country, are certainly very curious; but though described as three separate species, there seems every reason to believe, or rather, in fact no reason to doubt, but that they are all referable to one and the same butterfly—that before us; and that the opinion expressed by Mr. Edward Newman, in the "Entomological Magazine," volume ii., page 615, is correct; namely, that as they advance to the midland counties, "an evident change has taken place, the band of rust-coloured spots has become less bright; at Manchester these spots have left the upper wing almost entirely; at Castle Eden Dene, they are scarcely to be traced, and a black spot in the centre of the upper wing becomes fringed with white; in some specimens it is quite white; the butterfly then changes its name to Salmacis. We proceed farther northward, and the black pupil leaves the eyes on the under side, until at Edinburgh it is quite gone; then it is called Artaxerxes."

The plate is from specimens in my own cabinet.

GRIZZLED SKIPPER.

PLATE LXVI.

Hesperia Malvæ,	LEACH. CURTIS. DALMAN.
" "	ZETTERSTEDT.
" *Fritillum minor,*	FABRICIUS.
Papilio Malvæ,	LINNÆUS. LEWIN. HAWORTH.
" "	TURTON. HARRIS.
" *Alveolus,*	HUBNER.
" *Althææ,*	BORKHAUSEN, (*var.*)
" *Lavateræ,*	FABRICIUS. HAWORTH.
" "	JERMYN, (*var.*)
" *Malvæ minor,*	ESPER.
" *Sao,*	BERGSTRASSER, (*var.*)
" *Fritillum,*	LEWIN. (*var.*)
Pyrgus Alveolus,	HUBNER.
" *Malvæ,*	WESTWOOD.
Thymele Alveolus,	STEPHENS. DUNCAN. WOOD.
Syricthus Malvæ,	BOISDUVAL.

I HAVE taken this species in plenty at Pinhay Cliff, Devonshire, an "old familiar spot," near Lyme Regis, Dorsetshire; in Yorkshire, at Buttercrambe Moor, Allerthorpe Common, and Sutton-on-Derwent. Other localities for it are Sywell Wood, near Northampton, Barnwell and Ashton Wold, and the neighbourhood of Polebrook, Northamptonshire; Brighton, in Sussex; Raydon Wood, very abundantly, and other woods, near Colchester, Essex; Great Bedwyn and Sarum, Wiltshire; Yarmouth, in the Isle of Wight, and Bisterne, Hampshire; Bewdley, Worcestershire, as also in various parts of Kent, Hertford, Durham, Cambridge, Northumberland, and the south of Scotland.

It appears the end of May, and beginning of June.

The caterpillar feeds on the teazle, (*Dipsacus sylvestris.*)

In this very pretty little species, which is just about an inch in width, the fore wings are very dark brownish black, marked with about fourteen small white spots; the base is powdered with white, especially in the male. The fringe is wide and white, elegantly crossed with the black of the ground-colour. The hind wings are of the same colour, and marked in a similar manner, but the white spots are much smaller, fewer, and less distinct.

Underneath, the fore wings are paler, and the spots larger, clearer, and more run together in lines. The hind wings are principally of a neat brown colour, with large spots, one of them a wide short band from the front edge. The inner part of these wings is greyish black.

The caterpillar is green, with pale longitudinal stripes, the head black, and a yellow ring round the neck.

The chrysalis is wrapped up in folded leaves of the plant on which the larva feeds.

There is a not very common variety, which Fabricius and Lewin considered as a distinct species, in which, as Messieurs Westwood and Humphreys describe it, "there is a white oblong blotch on the middle of the fore wings towards the posterior margin, visible on both sides, which is frequently duplicated from the confluence of two contiguous spots. The white dots are also larger than in the typical individuals."

Mr. Stephens possesses a specimen with "one of the fore wings marked as in the variety, and the other in the type."

The plate is from specimens in my own collection, one of them the variety just spoken of.

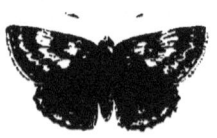 

DINGY SKIPPER.

PLATE LXVII.

Hesperia Tages,	FABRICIUS. LEACH. JERMYN.
Papilio Tages,	LINNÆUS. LEWIN. HAWORTH.
" "	HARRIS.
Thymele Tages,	FABRICIUS. STEPHENS. DUNCAN.
" "	WOOD.
Thanaos Tages,	BOISDUVAL.
Nisoniades Tages,	HUDNER. WESTWOOD.

I HAVE also captured this Skipper in plenty near Charmouth and Lyme Regis, Dorsetshire, and Devonshire; in Yorkshire at Buttercrambe Moor, and Sutton-on-Derwent. It is met with at Burnwell and Ashton Wold, and the neighbourhood of Polebrook, Northamptonshire; Brighton, in Sussex; and is very abundant in Raydon Wood, and Wintlesham Wood near Hadleigh, Essex; near Great Bedwyn and Sarum, Wiltshire; and, in fact, in most parts of England. In Ireland it is plentiful at Ardrahan, near Galway, as A. G. More, Esq. tells me. It is taken also in Scotland in different parts. In Wales on the sandhills near Llandudno.

It frequents for the most part wooded districts, both the woods themselves, and any places not very distant from them; open flowery pasture meadows, where, a "tenant for life," its sombre hue contrasts well with the yellow of the golden buttercup on which it alights.

This plain-coloured insect occurs at different periods, in May, June, and July.

The caterpillar feeds on the bird's-foot lotus, *(Lotus corniculatus,)* and the field eryngo.

The wings of this species expand to the width of about an inch and a quarter. The fore wings are blackish brown, with three cross-waved bands of grey, the middle one the widest, and most distinct. In some specimens the two colours contrast together much more markedly than in others. The margin is grey, edged on its inside by a black line, close within which is a linear row of small white dots. There are two white dots near the front edge towards the upper outer corner. The hind wings are blackish brown, with a few indistinct paler marks above the centre, beneath which is an indistinct row of paler dots, and then a linear row of white dots. The margin grey.

Underneath, the fore wings are of a uniform pale fulvous grey brown, with a small white dot near the outer corner on the front edge, and a line of dull white dots within the marginal line. The hind wings are of a similar colour, with a quadrant of dull white spots above the middle of the wing, towards the front edge, and a half line of others following inside the other line.

The colours in the female are brighter than in the male.

The caterpillar is bright green, the head brown, the back and sides with yellow stripes, dotted with black.

The figures are from specimens in my own cabinet.

LARGE SKIPPER.

PLATE LXVIII.

Hesperia Sylvanus,	FABRICIUS. VILLARS. GMELIN.
" "	OCHSENHEIMER. CURTIS.
Papilio Sylvanus,	HUBNER. DONOVAN. HAWORTH.
" "	HARRIS.
Pamphila Sylvanus,	FABRICIUS. STEPHENS. DUNCAN.
" "	WOOD. WESTWOOD.

THE habits of this kindred species are similar to those of the next but one, and "a life in the woods for me" will be the exclamation of every lover of nature who seeks it or any other butterfly in the calm and peaceful scenes where it is to be met with—the ridings and paths in woods, or their borders, sheltered lanes, etc.

It is common in most parts of the country; in Yorkshire at Sutton-on-Derwent, and a variety of places. Also near Brighton, Bisterne, Cirencester, Bromsgrove, Charmouth, Shanklin, Llandudno, etc. Near Falmouth it is in general scarce, but was plentiful in the year 1850, in a lane leading to Mr. Jagoe's farm, as W. P. Cocks, Esq., writes me word.

The perfect insect appears at the end of May, or beginning of June, and also at the end of July, or beginning of August.

This species measures rather over an inch and a quarter across the wings. The fore wings are tawny orange, with a large black vein following the front edge, and an oblique bar near the middle, from the end of which the black veins diverge; the outer part is tawny brown, two spots detached from the orange near the tip, breaking the line of the latter; the margin paler, edged inside with black. The hind wings are of the same general ground-colours, with a waved bar of spots of a brighter tint than the rest.

Underneath, the fore wings are marked as above, but much duller and less distinctly: the brown is tinged with green in some lights.

The hind wings have also a greenish cast, the spots shewing through, and the lower corner and margin orange.

The female is without the black bar or the fore wings; their base is also fulvous brown as is the margin, so that the orange forms a wide irregularly bar across the centre of these wings. The hind wings are tawny brown, with a waved bar of orange; the margin lighter, edged with black interiorly.

Underneath the colours shew through, but duller; the hind wings are principally of a greenish cast, the lower corner orange, and the bar the same, but paler.

The caterpillar has two white patches on the under side of the tenth and eleventh segments.

The figures are from specimens in my own collection.

SILVER-SPOTTED SKIPPER.

PEARL SKIPPER.

PLATE LXIX.

Hesperia Comma,	FABRICIUS. OCHSENHEIMER. CURTIS.
" "	BOISDUVAL. ZETTERSTEDT.
" *Sylvanus*, (*female*,)	JERMYN.
Papilio Comma,	LINNÆUS. HAWORTH. LEWIN.
" "	DONOVAN. HUBNER.
Pamphila Comma,	FABRICIUS. STEPHENS. DUNCAN.
" "	WOOD. WESTWOOD.
Augiades Comma,	HUBNER.

"YA!" said Suttum to Mr. Layard, as he waded his mare through the flowers of a thousand hues, with which the vast plains of Mesopotamia are carpeted, and the serene air perfumed, "Ya! what do the dwellers in cities know of true happiness; they have never seen grass or flowers! What worldly delight has God given to us equal to this? It is the only thing worth living for." Poor Suttum! he has since, it appears, fallen in a foray, but if he has a monument, let the epitaph inscribed upon it be, that he was a naturalist, and had a fellow feeling with every other lover of nature.

This species is plentiful near Newmarket, and at Gogmagog Park, near Cambridge, and Mr. Dale records the neighbourhood of Hull as another locality for it; Barnwell and Ashton Wold, and the neighbourhood of Polebrook, Northamptonshire; near Dover, Kent; Brough, in Yorkshire; Old Sarum, Wiltshire; Croydon, Surrey; Lewes and Brighton, Sussex, are also its habitats.

It is taken in the end of July, and the beginning and middle of August.

The caterpillar "feeds on the various-coloured coronilla, (*Coronilla varia*,) on the continent."

This Skipper measures an inch and a half across the wings. The fore wings, which are hollowed on their front edge, are dark tawny orange on their inner portion, the remainder greenish brown, with some light orange spots towards the tip; there is a black oblong bar near the centre; the margin pale yellowish fulvous, boundly darkly by the ground-colour. The hind wings are a mixture of orange fulvous and brown; the margin the same as in the fore wings.

Underneath, the fore wings are marbled with dark greenish at the tips, yellowish brown, orange, black from the base, and some silvery spots. The hind wings are brassy green, with a waved row of six silvery spots, and three others, the lower ones often in conjunction, near the base: the inner lower corner is orange.

The female has the fore wings, which are more hollowed on the edge than in the male, dark tawny orange on their upper inner portion, the remainder greenish brown, with several light orange spots in an irregular waved bar, the margin cream-colour, as in the male. The hind wings are of a deep greenish brown round the edge, which is margined with cream-colour, the remainder is the same, but with a cast of orange in some lights, and a waved bar of dark orange following the upper part of the outer margin.

Underneath, she is marked similarly with the male, but the fore wings are much more handsomely and distinctly marbled, and all the colours brighter, and the hind wings darker green, on which the spots are much brighter and more distinct.

The caterpillar is of an obscure green colour, marked with dull red; the head black, the neck with a white collar, and a row of black dots on the back and sides. It has two white spots on the under side of the tenth and eleventh segments.

The chrysalis is elongated, and of a cylindrical shape.

The plate is from specimens in my own collection.

SMALL SKIPPER.

PLATE LXX.

Hesperia Linea.	FABRICIUS. OCHSENHEIMER.
" "	LEACH. CURTIS. BOISDUVAL.
Papilio Linea,	HAWORTH. DONOVAN. HARRIS.
" *Thaumas*,	ESPER. LEWIN. STEWART.
" *Comma*,	BARBUT.
" *Flavus*,	MULLER.
Pamphila Linea,	FABRICIUS. STEPHENS. WOOD.
" "	DUNCAN. WESTWOOD.
Thymelinus Linea,	HUBNER.

THIS active and lively little species is to be found in woods, or along their borders, and "for whom is the forest so pleasant and gay" as for the quiet insect-hunter, who there sees it in the gladsome sunshine, flitting from flower to flower with its burnished wings?

It is rather uncommon in Great Bedwyn and Sarum, Wiltshire, as J. W. Lukis, Esq. informs me, but in most parts of England is very abundant, as at Brighton, Charmouth, Bisterne, Sutton-on-Derwent, Buttercrambe Moor, Sandal Beat, near Doncaster, Shanklin, Looe, Ranmore Common, near Dorking, Surrey, etc.

It frequents open places in and near woods.

The fly appears towards the end of June, in the beginning and middle of July, and the middle of August.

The caterpillar feeds on the mountain air grass, and other grasses.

This fly a little exceeds an inch in the expanse of its wings. The fore wings are bright bronzed fulvous; the margin paler, edged interiorly by a line of black; the base dusky; there is a narrow oblique black bar near the centre. The hind wings are of the same ground-colour and margin; the upper edge widely bounded with black.

Underneath, all the central part of the fore wings is also fulvous, the base and lower edge dusky, the outer corner greenish. The hind wings are also of a bronzed greenish hue, the lower inside corner orange.

The female wants the black bar, and there is no dusky colour about the base. In other respects she resembles the male.

The caterpillar, which is of solitary habits, is of a deep green colour, with a dark line down the back, and two whitish lines along the sides, edged with black. It has two white patches on the under side of the tenth and eleventh segments.

The chrysalis is of a green colour, and is enclosed in a slight cocoon.

The engraving is from specimens in my own cabinet.

LULWORTH SKIPPER.

PLATE LXXI.

Hesperia Acteon,	OCHSENHEIMER. CURTIS.
Papilio Acteon,	ESPER. HUBNER
Pamphila Acteon,	STEPHENS. WOOD. DUNCAN.
" "	WESTWOOD.
Thymelinus Acteon,	HUBNER.

IN company, some years ago, with my friend J. C. Dale, Esq., late High Sheriff of Dorset, I formerly captured this, then newly by him discovered insect, I mean as a British one, in plenty at Lulworth Cove, Lulworth, Dorsetshire—a charming scene, where you will be fain to wish that you could for ever watch the glorious ocean, dashing up from its dark depth against the steep cliffs, which there present an aspect of the utmost seclusion and the most lonely retirement. Wild must all around be in winter, but this small butterfly rejoices in the settled summer, more fortunate than some of its class, who are tempted out to woo the "beautiful spring;" often their reception is cold and chilling, and their day-dream of happiness is blighted like the contemporary delicate flower that has peered out too soon from its sheltered nook, and must again hide its head for a season till the skies are more propitious and the sun shall shine undisturbed upon it. Now it is not to be seen there, though it is still to be found at the burning cliff, near Weymouth, where my friend the Rev. Francis Lockey, of Swanswick Cottage, near Bath, has taken it in plenty; Mr. Humphreys, also at Shenston, near Lichfield.

This butterfly appears in July and August.

The date of the appearance of the caterpillar is in June and July. It feeds on the reed, (*Arundo phragmites,*) and the *Aurundo* (*Calamogrostis*) *epigejos,* rolling up the leaf by white silken cords across; sometimes two larvæ are found in one leaf.

In this small species the expansion of the wings is about an inch; the fore wings are of a bronze brown colour, with an orange brown dash, and a short bar of the same, also a straight black horizontal line in the middle of the wing; the fringe darker, edged on the inside with an evanescent dark line. The hind wings are of the same general bronze brown colour, the dark line near the edge more distinct and wider.

Underneath, the wings are more of an orange cast, but in some lights it is not diffused, but the ground-colour appears.

"The disc of the fore wings in the female is more tawny orange; beyond the dark extremity of the discoidal cell is a curved series of six or seven orange spots, divided from each other by the veins of the wings. The under side has a pearly ochre lustre; a large orange patch on the fore wings extending to the tip of the discoidal cell, where the pale row of spots again appears, but more obscurely, and an oblique portion of the inner edge of the hind wings yellowish orange."

The caterpillar is tapered towards the head and the tail, the latter being large, and as it were almost separated from the body. It is of a pale greyish green colour, with a darker line along the back, edged with a slender pale yellow streak on either side, and enclosing a pale line along its middle; a narrow yellowish line runs above on the side, and a broader one below. The two lines on the back reach to the middle of the head, and at the tail is narrowly ended with pale yellow. The head is at first brown, afterwards paler, with two distinct yellowish lines edged on the outside with brown, greenish in older ones, with shorter and paler lines, without darker edges. There are two white patches on the under side of the tenth and eleventh segments.

The plate is from specimens in my own collection.

SPOTTED SKIPPER.

CHEQUERED SKIPPER.

PLATE LXXII.

Hesperia Paniscus,	FABRICIUS. OCHSENHEIMER. LEACH.
" "	JERMYN. CURTIS.
Papilio Paniscus,	DONOVAN. HAWORTH.
" *Brontes,*	HUBNER.
" *Sylvius,*	VILLARS.
Pamphila Paniscus,	FABRICIUS. STEPHENS. WOOD.
" "	DUNCAN.
Steropes Paniscus,	BOISDUVAL.
Cyclopides Paniscus,	HUBNER. WESTWOOD.

This is a very local species, though where it does occur it is found in abundance.

Sywell Wood, near Northampton, Milton, Rockingham Forest, Monks Wood, and Castor Haglands Wood, near Peterborough, Barnwell, Ashton Wold, and the neighbourhood of Polebrook, Northamptonshire; White Wood and Gamlingay, Cambridgeshire; near Dartmoor, Devonshire; Linwood, near Market Rasen, Lincolnshire; Newark, Notts.; Clapham Park Wood and Luton, Bedfordshire, are localities for it. I have heard that in one of these places the "Lord of the Manor" has forbid the "free warren" and "free entry" of the entomologist, but I am unwilling to believe that any such interference with the "liberty of the subject" has been perpetrated.

The end of May or beginning of June is the time when the Spotted Skipper first appears, and in July it is still to be seen.

The caterpillar feeds on the greater plantain, (*Plantago major,*) and the crested dog's-tail grass, (*Cynosurus cristatus.*)

This Skipper is about an inch and a quarter in the expansion of the wings. The fore ones are of a rich dark brown ground-colour, spotted horizontally and transversely, with large black spots, and a wide black border on the outer side, with a faint row of dull small orange dots following the margin, which is dull pale fulvous. The transverse bar of spots has two, which would otherwise complete it, pushed "out of their propriety," towards the outside edge. The hind wings, which are of the same ground-colour, have some large round similar orange spots on their middle, and an irregular row of smaller ones outside them; the margin, pale dull fulvous.

Underneath, the ground-colour is tawny yellow on the fore wings, the dark marks seen obscurely and partially through. The lower wings are of a dull greenish cast, some of the spots shewing through of a dull pale buff colour, and others being added near the front edge, and the row near the border being larger. The antennæ on the lower side are bright orange.

The female resembles the male.

The caterpillar is dark brown on the back, with two yellow stripes on the sides; the head is black, and there is an orange ring round the neck.

The chrysalis is of a dull grey colour.

The plate is from specimens in the cabinet of the Rev. William Bree.

Small as this butterfly is, and insignificant indeed as every insect may by some be thought to be, unworthy of serious attention, yet, when we come to regard it with reference to other creatures, we shall see reason for thinking far otherwise; for, to say nothing of its wonderful organization and wonderful beauty, it holds, in truth, a comparatively high place in creation. Thus Ehrenberg informs us that there are some of the animalculæ so minute, that five hundred millions of them might be contained in one of the drops of water which do, we know, actually contain such numbers of strange and different species of living beings. The microscope reveals to us wonders in one way equally great with those which the telescope brings home through the eye to the mind in another, even though the latter, with its still limited power, shews us that in all probability the sun is but the centre of one system, and that there may be others, perhaps countless others, in comparison with some even of which ours may be insignificant, revolving each in their prescribed orbits in the regions of infinite space. There is indeed a "Music of the spheres," which is heard by the soul alone, and it sings the power of HIM who is the Eternal, the Almighty, and with its silent voice invites us to join in its harmony with the unspoken and unutterable language of the heart. "And these are but parts of HIS ways, but how little a portion is heard of Him, but the thunder of His power who can understand?" In the contemplation of the limited portion of the works "which God created and made" that it comes within the bounds of our knowledge or of our capacity in some small degree to comprehend, the mind is lost in admiration, the soul appalled with awful reverence.

Every one of those minute, and to the eye invisible creatures I have alluded to, which yet again may be gigantic compared with others which even the aids of science will not enable us to discover, has all its internal organization complete, and adapted in the most absolutely perfect way to all its requirements, and is even able, as has been proved, to impart animal heat to the fluid in which it lives. How astonishingly small then must each separate part of each be, all acting in as harmonious co-operation as any of those of the higher orders of earthly being! Every individual of them too has, so to speak, mental capacities, by which the actions of their bodies are unerringly ordered and directed. Yet, "known unto God are all His works from the beginning of the world," and every separate action of every separate animalcula is known to Him, both before its occurrence, and as afterwards registered, as well as every motion of every vast planet, and the history of each atom of its component parts!

I conclude my "History of British Butterflies" with the sentiment and in the words of an old writer, "The Majesty of God appears no less in small than in great, and as it exceedeth human sense in the immense greatness of the universe, so also doth it in the smallness of the parts thereof."

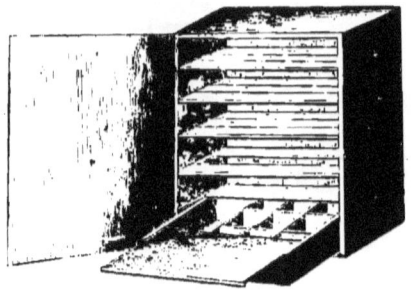

Setting Case.

Store Box.

Breeding Case.

Pocket Box.

Relaxing Jar.

Chloroform Bottle.

Digger.

Sugar Tin.

Insect on Turned Wood.

Thread Spool.

Box for Pill Boxes.

Round Net.

Clap Net.

Emperor Net.

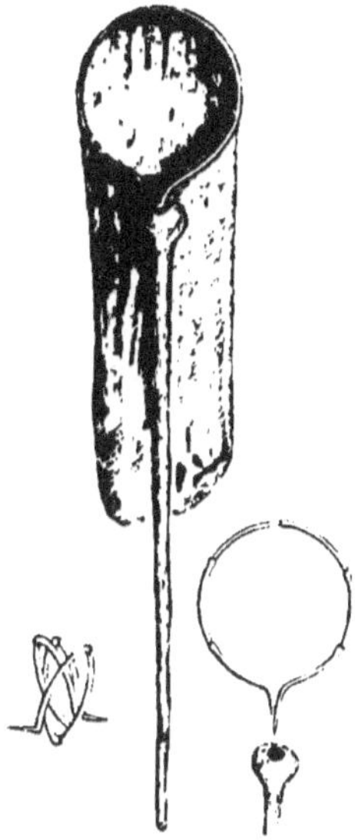

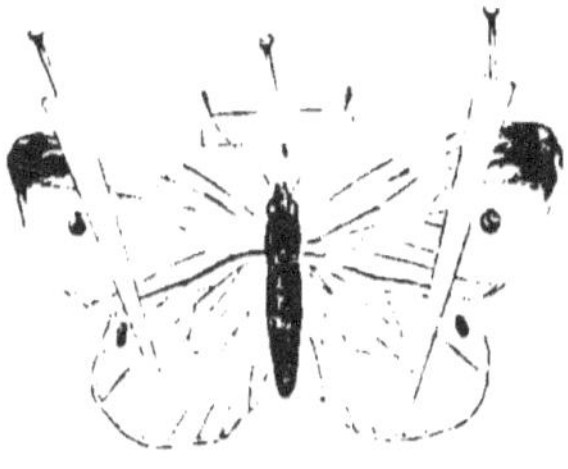

Insect with Single Braces.

Sweeping Net.

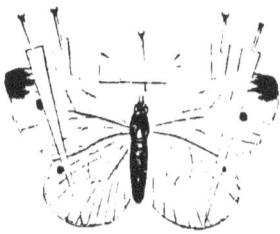

Insect with Double Braces.

Fly Extended.

APHORISMATA ENTOMOLOGICA.

"CAVENDO TUTUS."

Nothing can be done in Entomology without a good cabinet: this should be the foundation stone. Have it made of old oak or mahogany, either of these woods being well seasoned. It should be well and evenly corked, with good soft cork, and glazed with good glass; the glasses fitted in frames to take in and out. It should be made by some person who is in the habit of making them, for the mode of fitting the frames is not very easily explained on paper.

Let the drawers be about one foot nine inches long in front, one foot six inches in width, and two inches and a quarter in depth, on the outside. They must be carefully papered at the bottom and on the sides. This is always done in the first instance by the maker. There must be a ledge placed for camphor, but it need not go all round the inside of the drawer, as is generally the case; it will be quite sufficient to have it on one side. This had better be the front side, as then it is completely out of sight, and the drawer appears without any detriment. There need be no holes cut in the front of the ledge, for the scent of the camphor to pass through; the top of the ledge being left open affords abundant escape for it.

Keep the cabinet in the driest room in the house, and never let it be placed against an outer wall, but if possible against the part of a wall that is behind a fireplace in an adjoining room. Mould must be carefully avoided: it is thus totally prevented; but if otherwise suffered to appear, cannot be cleared away without some injury to the specimens, and will spread again unless thus checked. Keep the glasses on the drawers as much as possible, both on account of the mites, and also of the dust, which, if it settles upon the insects, must more or less damage their appearance. Take, however, the glasses off every now and then for a moment or two, or raise them, if ever so little; for the fresh air admitted will help to keep off mould and damp. See that the drawers of the cabinet run easily, otherwise the specimens will be shaken by the jarring every time it is put in or taken out, and the antennæ and bodies will be liable to be shaken off.

When, in time, or by accident, the paper gets discoloured or damaged, do not paper the drawer over again, but whitewash it, which has an excellent effect, both in making it whiter than it even was before, filling up all pin-holes, destroying mites, if any, and stuffing up all those chinks and crevices where they resort, and probably is a preventive of decay and injury generally. Common whitening will do very well, and it may be laid on with a common paint brush of a middle size. This, however, must be carefully done, as the more smoothly and evenly it is washed over the drawer, the better of course it will look. It must not be laid on too thick. A thin coating will hide most blemishes when it is dry, but if the lines, of which presently, are intended to be re-arranged they had better be first rubbed out with a piece of Indian-rubber. Flake-white used to be considered the best material to make the wash of, but it is rather expensive, and it will be found that precipitated chalk, which is sold for a penny an ounce at the druggists', will answer the purpose as well, or better. Six pennyworth of it will be enough for twenty drawers. It should be placed in a common small basin, and boiling or cold water poured upon it, so as just to cover it. Some good gum arabic, dissolved, should then be added, as size, to make it keep without rubbing off. A tablespoonful or two will be enough for two or three ounces. This is also to be had ready made at the druggists', and costs a mere trifle. More water may be added from time to time, as required. Experience will soon teach the right proportions of each of the ingredients. The paste on the paper is a never-failing supply of food for the mites, and the lime in the chalk is an excellent corrective.

To expel or destroy mites, invert the drawer, and place it, the glass frame having been taken off, over a sheet of blotting-paper well saturated with liquid naptha for an hour or two. It is also recommended to leave a few small globules of quicksilver loose in the drawers.

Cork is the thin to line the drawers with, but the following is a cheap substitute, and easily procured and applied:—

Two-thirds of the best bees-wax; one-third of the best resin; to which, in this climate, not being a very hot one, a little tallow may be added.

Renew camphor in the drawers every three months.

If any mould arises on the antennæ, wings, or bodies of any specimens, a little cajeput oil will be found the best possible remedy. It must be applied with a very small camel's hair brush. The best oil should be procured, and, if possible, direct from the Apothecaries' Hall. It will be found also most useful in thus restoring beetles, and has a relaxing effect at the same time upon the antennæ and legs of the smaller ones, so that their setting can be easily improved if necessary.

If the cabinet can only be kept against an outside wall, let it, if possible, be a wall with a south aspect.

In placing the insects in rows in the cabinet, draw double lines between each row; it has a much better effect. Use great neatness in drawing the lines, which should be made with a hard and very fine-pointed pencil. Put the insects very straight in each row, or the collection will never look well.

Leave space, in all cases where you have not already got a series, for four specimens. Of those species which are liable to vary much, a whole row should be kept.

"PRACTICE MAKES PERFECT."

It would at first sight almost seem like "putting the cart before the horse," to give, as I have done, instructions respecting the cabinet, before I have said a word as to the mode of capturing the insects which are desired to be placed in it. Such however, is by no means the case: there is no doubt whatever as to the capture of a vast variety of species, but if there is not provided, beforehand, a suitable receptacle for them, they will in all probability be wasted; and it would have been far better in that case never to have captured them at all, but that they should have been left flitting along the side of the hedge, or over the open meadow, or in the paths and rides in the woods.

Now, therefore, as to the 'modus operandi.' A vast variety of different kinds of nets have been invented and described, but depend upon it, that there is none better than, or so good as the common one, of which an engraving will accompany these remarks. It is made to take to pieces, and put up in the pocket of your coat. Those who in their younger days have known the kind of engine that is used when you go out on a dark winter night, with a large "bat-fowling net" in their hand, and a lantern with the means for lighting it in their pocket, will understand at a glance the whole art. More need not be said by way of description; the figure will explain itself to every one that is desirous to learn. This, I say, is the best kind of net. The bottom part should be turned up, and run on a piece of tape tied to the handles on each side, forming a kind of bag, to stop the captured insects from falling down and out of the net if held upwards at the time.

The second best, and indeed the only other that I at all recommend, is also best to be understood by means of an engraving, which is accordingly given; but I may mention that though less fitted 'ad captandum' than the one just before mentioned, the present has an advantage, or rather some advantages over the other, though on the whole in point of usefulness inferior to it. First, it folds up into a very small compass. Secondly, it is shut up or put together in a very brief space of time; which when you see three Convolvulus Hawk Moths at once, as I have done, is a considerable desideratum. Thirdly, the stick to which the round net is affixed, is neither more nor less than a common walking-stick, and useful accordingly in more ways than one: add to which, that even when fitted to the net, or rather the net to it, it will often be found very serviceable in "beating about the bush," as you walk along; and if any "scarce article" flies out from its concealment, you instantly reverse the state of affairs, and capture the insect, or—miss it.

It is then very serviceable, at all events as a 'pis aller,' though inferior, on the whole, to the first-named net. As to the forceps 'et id genus omne,' I exclude them altogether from my vocabulary of entomological apparatus. A common fishing-jacket, with two large and two small pockets outside, is the best kind of coat.

"FIRST CATCH YOUR HARE."

"BLESSINGS on the man that first invented sleep," said Sancho Panza, and every housemaid must surely almost as much praise the memory of him who was the first discoverer of lucifer matches. The old tinder-box and its concomitants must be most unpleasingly associated in their minds with the idea of a cold dark winter morning, and the almost impossibility there was in the olden days, or rather nights, to "strike a light." Now, thanks to the benign inventor of the lucifer match, whose sole thought must have been doubtless for the unhappy females spoken of, all difficulty is at once and for ever removed, and if they only "keep their powder dry," it will give them the privilege of lying in bed half an hour longer than they otherwise would be able to do, with the positive certainty of being in time even for an early breakfast.

There are, however, certain individuals who must lament the loss of the old-fashioned brimstone match: for as it is "an ill wind that blows nobody good," so it must be one that comes from all points of the compass at once that does no one any harm. The entomologists are they to whom I allude, for it is now next to impossible, at least I find it so, to procure even a halfpenny worth of the article once so universally in vogue. But, 'cui bono?' what are the brimstone matches for? The next chapter will tell you; containing, as it will do, a "dissertation on boxing," and that of a pugnacious kind too, albeit altogether peaceful. The solution of the riddle will there be given.

"PUGNO, PUGNAS, PUGNAVI."

IF you can construe the above without conjugating it, somewhat after the same 'ideal' as "malo, malo, malo, malo, quam vivere malo, malo," you will probably be able to understand somewhat of the allusion in the last chapter.

As I have already recommended the use of but few kinds of nets, so I would give similar advice as to boxes. You must, however, have some of different sorts. They may be made of oak, mahogany, or deal; or one or two of them, of which presently, of tin. First of all, in further carrying out the principle laid down as to the necessity of a previous state of preparation, you must have a large one, say two feet long, one foot wide, and three inches deep, on the outside. Of course it must be lined with good soft cork, and papered, and, if thought advisable, from time to time whitewashed over, in the way described for the cabinet. A little camphor should be securely fixed in one of its corners, and be regularly renewed from time to time; or a piece of cotton wool, to be saturated now and then with cajeput oil. It should have a lid projecting above the inside edges of the lower side all round, to keep the dust from penetrating through the interstice when the box is shut together. This box is for carrying about with you from place to place, as a temporary receptacle for your captures when preserved, or any species that you may procure by gift, exchange, or purchase. Of course it must be shaken as little as possible, and, when you are travelling, should be carefully packed in your "portmantel." If you "do business in a large way," you will require several of these boxes.

The next box to be procured, and to be now described, is of much smaller dimensions, being what is commonly called the pocket box. It may be made about six inches long, four wide, and two deep; but on the same principle that you "cut your coat according to your cloth," so you can have your box made larger or smaller according to the size of your pocket. Now, let this box be made of tin; and as to the mode of making it, I have to give myself credit for, in the words of my namesake, Miss Edgeworth's Francisco, "a discovery! a discovery! which it concerns all" entomologists "to know!" as follows:—

Let this box, I say, which is to take out with you when you go collecting, be made of tin, and be of the dimensions just given, or as nearly so as may be most convenient to yourself. Have it made to open as shewn in the plate, not in the middle, as these boxes generally are, but nearer to the top, so as to have only one side, the bottom one, lined with cork, which should be papered or whitewashed over, for the reception of recent captures. Inside the lid, have a piece of perforated zinc, which you can obtain at any good ironmonger's shop—fine wire network would do, but that it is liable to rust, especially under the circumstances about to be narrated. Videlicet; this piece of metallic gauze being fixed on a little hinge or hinges at the inner edge of the lid, is to be made to open out, or shut in, at pleasure. Between it and the lid, place a flat piece of sponge, and when you are going out collecting, dip the top of the box, thus containing the sponge between the actual lid and the "fly leaf" of zinc, in water. If it should become dry, or rather so, which will naturally be the case in the hot times of the year when for the most part you go out collecting, all you have to do is to dip it again in the first stream of water you come to, which will probably not be again required to be done. The effect is this; instead of your insects, even if ever so small, being dried up by the time you return home, so as to be incapable of being set until you have been at the additional trouble of relaxing them, they are as fresh as at the moment they were first captured; and if you have not time to extend them all that night, by again moistening the sponge, and keeping them in this, so made, relaxing box, you will find them still pliant the following morning. 'Intelligis-ne?'

The mention of the small moths brings me to the third kind of box required. This, or rather these, for you should have two or three, or more of them, is, or are also to be made of tin; and if lacquered or japanned their appearance will be rather more neat. They are to be made of small size, say four inches long by three wide, and about three quarters of an inch in depth. Fill them with small pill boxes, each with the lid perforated with a number of pin holes to admit the air, and the fumes of sulphur when applied, as about to be explained.

Instead of putting any minute moths you may catch into the pocket box, put one into each pill box, enclosing them by holding the box, with the lid off, against the part of the net where you have confined them for the purpose, and then, "stealing a march" upon them, putting the lid on again; all this a little practice will explain. Thus you have not touched them at all: and on your arrival at home, or wherever you want to see them, pile the pill boxes under a tumbler near the edge of a table, only do not let it be your best one; light two or three brimstone matches; draw part of the glass over the edge of the table; hold the matches underneath, so that the fumes of the brimstone can ascend into the glass, taking care not to touch the glass with the light, or it will be cracked; and as soon as you see, or rather when you can see nothing in the glass for the smoke, replace it entirely on the table, to confine the vapour, and in a few seconds all the moths will be apparently dead, and by leaving them there for a little while they will become entirely so. Then is the time to set them, and to set them well, uninjured in the smallest degree by the touch of your hand.

These pill boxes will often be found very useful for bringing home the smaller caterpillars in; and for the larger ones, any small boxes will do. I have found the round-turned lucifer match-boxes to answer admirably for this or any other kindred purpose. But it is getting on towards midnight, and I must for the present conclude.

"A TRAP TO CATCH A MOONBEAM."

Another mode of capturing moths—"unde a quo abi redeo'—is by means of a light—to which, in the dusk of the evening, they are attracted I proceed to give two or three different methods of procedure. One plan, of primitive simplicity, and which was adopted by us at school, was to place a candle near an open window; tie a long string to the handle of the frame—they were old-fashioned lattice ones—get comfortably into bed, and when a moth made his 'entree' pull the window instantly to, thus securing him within.

The following is a much more elaborate method, invented and adopted by my friend, the Rev. Francis Lockey, of Swanswick Cottage, near Bath, to whom, knowing how successful he had formerly been in its practical operation, I wrote for particulars, which he has obligingly furnished me with, as follows:—

"Many years have elapsed since I have engaged in entomological pursuits, but I most readily reply to your inquiry as to the mode adopted of moth-capturing at night.

The windows of my study, partly for this purpose, and partly to secure more equable temperature, are double. Each window consists of two pairs of sashes, with a light mullion of wood between the *interior* pair. To make it more clear I have sketched an horizontal section, wherein A represents part of the thickness of the wall of the apartment; B and C are sections of the outer window-frame; and D and E are sections of the inner window-frame. The outer window opens outwards, and the inner window inwards, as indicated by the arrows near C and E. At F and G two small rings are fixed, and to these are fastened cat-gut strings, marked by the dotted lines which pass through the mullion H, and enable one to open and close these exterior windows without the inconvenience of opening the interior. I ought to explain that when the cat-gut strings are set free, (in the room,) the windows open freely under the influence of a weight and lever, not shewn in the sketch, and in fact concealed in a trunk or box in the thickness of the wall at A. The space between the two windows is about six or seven inches.

Evening having arrived, the outer windows were allowed to open, as in the sketch Fig. 2, and a lamp L placed on a shelf within the inner window. This lamp was an Argand, and moreover furnished with a powerful Parobolic Reflector, about sixteen inches in diameter.

The moths usually announced themselves by striking against the interior window, D or E. The cat-gut string at H was then pulled, (H F, H G,) and the capture being thus enclosed in the space between the windows are readily reduced to a closer captivity either for examination or possession. Only a few occasions the window had been perfectly besieged by moths, and 'at one haul' I think it was at the beginning of one July, some hundreds of moths were enclosed.

To effect this second or closer capture it was of course necessary to use only one hand, which was armed therefore with a gauzed forceps, or, which I found more convenient, a large bell wine glass fitted up for the purpose in the following manner:—The glass B having had its foot broken off, is cemented to a box-wood handle, A. Opposite to A a slot is made about an inch in length, and wide enough to let the wire frame, presently mentioned, traverse it freely.

A wire frame, A, D, C, E, is formed, following the outline of the glass and handle, and bearing at the rectangular end C a disc of card-board, blackened. This disc, which is rather larger in diameter than the mouth of the glass, is attached to the straight part of the wire by a sort of continuous staple, formed by glueing over it a strip of paper. The disc therefore is moveable about the line C as an axis, whilst the part of the wire moveable in the slot A enables one readily to remove the disc from the mouth of the glass, as in Fig. 3; or, when the glass has been placed over the captive, to close it as in Fig. 4.

If the capture was intended to be retained, the closed glass was removed to a small stand, beneath a hole in which was a bottle containing the very strongest ammonia, or other more effectual vapour destructive of life.

The thin cardboard disc being now slipped aside, the insect was exposed to the vapour. In a short time, all consciousness having been destroyed, it seemed the safer plan to make sure of the extinction of life.

You will perceive that by this system the insect was never touched by the fingers, and its perfection was unimpaired."

"IN EXTENSO."

IF one ever thinks at all about the various facts with which we are necessarily conversant in every-day life, it can hardly fail to occur to the mind, that not only the origin of many of the most useful of the "appliances and means" with which we are even the most familiar, is lost in the mists of antiquity, but that the very names of the discoverers and inventors of the most useful and beneficial sciences and arts are for ever buried in oblivion, if indeed it was at any time their lot to rise from the obscurity which too often shrouds the most meritorious and deserving benefactors of the human race.

Who then was the inventor of the mode of setting insects that I am about to mention and explain, I am utterly unable to say, and perhaps no one may now know. Possibly the "ephemeral" nature of the subject may have been thought to have imparted a derived unworthiness of fame to the individual; but the society of entomologists should not be, and it is to be hoped will not be, blind to his merits, and, at all events, so far as the pages of the "Aphorismata Entomologica" can throw lustre on whomever or whatever they treat of, the invention shall be duly chronicled, even though the inventor's own name cannot be handed down with it to future admiring generations, but he must remain, perhaps for ever, "the Great Unknown."

On second thoughts, I think I will adopt the plan of beginning with the more simple mode, from which we shall advance, by one intermediate stage, to the more 'recherc' one. The analytical and the synthetical method has each its separate advantages.

Now the old procedure was to extend the butterfly or moth with one card brace impressed over each wing. Of course there are very many different degrees of excellence in the exercise of the art of doing this, the great thing being to have the wings exactly corresponding in extension on each side, and also, if I may so express myself, flat, in a sloping direction, namely, sloping down from the body of the insect to the surface on which it might be placed. This is accomplished by slanting the brace when struck into the cork. In very many instances it answers well, and the 'tout ensemble' is good; but on the other hand, in no means a few cases, and this is not to be guarded against, the wings become hollowed down in the middle, causing at the same time a turning up of their edges, the effect of which, to the entomological eye, is exceedingly bad. For a sample of this mode of extension see the plate. To remedy the defect caused by the practice of this mode, it occurred to some ingenious person—the date of the discovery I do not know—to support the wing from underneath by means of a supplemental card brace, before pressing it down with the one from above, this latter being placed more towards the edge than the lower one, thus pressing down the outside of the wing from the middle part which is raised by the brace beneath; the effect of the whole being to give it an elegant and graceful rounded appearance. For an illustration of this, also, see the plate.

But here "I maun premeese" that in both these ways of extending lepidopterous insects, as well as in the third, to be yet discoursed of, the great thing, at least it is a 'sine quâ non,' is to have the pin put perfectly straight into the insect, or if it is not, the fly, though ever so well set, will not look well; and it should also be put a good way through, for a reason to be hereafter mentioned when treating of the subject of entomological pins. "Item,"—The pins used for the card braces should be either the long lace ones, which hold tight in the cork, and are very sharp pointed, or the large common ones, which I find even still better.

"Au revoir." The third, the "grand climacteric," is to be essayed as follows: —If you can turn, I mean turn wood, in a lathe, yourself, you can make the required apparatus for yourself; but, if not, you can readily have it made for you by any turner: see the plate.

There are different modes of turning these pieces of wood, but the choice of these you must leave to the turner; suffice it to say, that if turned in the first instance in an oblong oval shape, each of these may be cut into four pieces of the proper sort for the extending woods. There is also a way of making them without employing a turner at all. Go into a carpenter's shop, or into your own if you have one, and plane down flat strips of wood round on each side: this in the various stages is depicted in the plate. Then, with a "plough" plane, run a groove along the centre of it, next cut it into suitable lengths, and finally with a "spoke-shave," shade it off down to one side—the one to be next you when setting the insect—cutting off all the edges square, as presently to be described in the case of the turned pieces of a similar shape. They must moreover, in whichever way they are made, be of different sizes, say two inches long by one and a half wide, three by two wide, four by two and a half, and so on.

They may be made of any wood, but common deal is the best, for several reasons: first, it is the cheapest; secondly, it is the most easily to be procured; thirdly, it is soft, and will admit of a pin being easily stuck into it; and fourthly, it is, though perfectly smooth when turned, rather rough when sawn through, which as presently shewn will have to be done. The advantage of the last-named particular is, that the threads, when wound round the insect, have a hold and do not slip. If the wood is hard, fine-grained, and smooth, the edges must have little notches filed or cut into them, all the way round, to hold the threads. So also as to a pin being easily stuck into it: this has an advantage, or, it may be, a double advantage; as thus:—If the wood be very soft, and the pin a strong one, you can extend the insect on the rounded wood with ordinary card braces, as if on a piece of cork, and the wings will thus, when dried, preserve the curved appearance so much admired by English collectors. Also, if threads be used, you will find that unless very great care be used to wind the thread round as lightly as possible, it will leave marks of its bandage on the wings, and even with the greatest care it is hardly possible with some species to avoid this; as for instance with the Whites and the Blues. To avoid this, therefore, some collectors first fix the wings on the (deal) wood with a pin in each of the two fore ones to the desired extent, and then by placing a small piece of silver paper on them the thread is wound round and round them without any detriment.

But to return; the pieces of wood when first turned, will of course be rounded down to the edge, but you will find that, if left so, it will be very difficult to take them up from any flat surface on which they may happen to stand; and to remedy this defect, you must have them cut off with a saw or a chisel on the other three sides besides that on which they have already been dissevered from the oblong turned piece; also, a narrow piece will have to be cut out transversely in the middle of each, to about half their depth, and a proportionate strip of cork be glued at the bottom of the cleft, filling it up half way again, and be shaved off to the shape of the wood on the side towards you: on this the insect is to be placed, and placed upright, with the pin straight—perfectly straight, through it, that is to say, through the middle of the thorax: otherwise, let me again and again impress on you, the insect will never look well, no, not though ever so well set.

Let thus much then suffice for the ground work; now for the further illustration, by way of clue to the "net-y-covered" labyrinth.

"YORKSHIRE EXTENSIONS."

THE thread to be procured is some that is used in "crotchet work," and commonly known by the name of "Moravian thread." It is not generally obtainable at a draper's, but is to be sought for at one of those shops in which Berlin wool and the materials for ladies' fancy work are sold. Thread of the same kind, or at least one that appears to be so, being in fact the single "strands" that compose a thread before they are twisted together into one, is to be procured in the manufacturing districts, being there known by the name of "Cop thread." If this is wound in a single thread on a spool, place the spool on a bar of wire or wood, as in the plate, and the means for extending a lepidopterous insect will, so far, be always ready to your hand. If it be not wound in a single thread, but two or three together, not indeed twisted, but singly, so to speak, you will probably have some kind female friend, some "neat-handed Phyllis," who will unravel the difficulty for you, and accomplish the work to your entire satisfaction. This I speak of the Moravian thread, but if you cannot have any wound singly, "French Embroidery Cotton," if equally fine, will be found equally excellent for the purpose.

Next, then, having fixed the pin which holds the insect—always, "Be it remembered," carefully and completely killed in the first instance of all—

straight in the cork in the groove, into which the body is just allowed to enter, holding the end of the cotton thread at the lower side, the one next you, of the wood, with your left hand, wind it once round the right hind wing of the insect; then, holding the thread round the lower part of the upper end of the wood, and also with the left hand, just, and only just, sufficiently tight to keep the wings in place, adjust the wings with the point of a large common thick pin, held in your right hand, to the desired extent, and then "lightly t(h)read" a sufficient number of times round in the same way, so as to keep all parts of the wings close to the shape of the wood; then, but not till then, completing, namely, first the right hand side—for otherwise if the threads be crossed and re-crossed there will be great danger, in taking them off, of breaking the antennæ, or in some way damaging the specimen—perform the same operation over the left hind wing.

Be, I say, very careful in again unwinding the thread, or woe betide the antennæ of your specimen, and with them will go its especial value in the eyes of the collector. The best way, however, is to "cut the Gordian knot," namely, cut the threads with a penknife against one side of the wood, or, better still, against both sides, and then the fly is at once taken off without further trouble. The whole spool of cotton costing only two or three pence, the time that would be required for saving the thread is gained, and amply repays the cost of a new one; and indeed, even if the different short threads were to be preserved, you would find that they cannot well be kept without becoming entangled together.

"ANOTHER MODE."

THE turned pieces of wood answer most admirably when you are quietly stationed at home, or fixed for a time sufficient in any other place, but they by no means suit the "locomotive department," or, rather, it does not suit them. To meet this difficulty, you can have recourse to either of the three following expedients:—First, when out collecting, and especially if you catch a large number of specimens, do not attempt to set any of your insects at all: on the whole I recommend this plan. You will find by the adoption of the method of relaxing hereinafter mentioned, that insects may be set quite as well, one may almost say even better, than when quite fresh caught. By keeping them also till the winter, or against a "rainy day," you will have the "Use of sunshine" for collecting, and be able to perform that afterwards leisurely which cannot be well done in haste. Secondly, by having your extending boards, of which more anon, narrow in width, you can tie a number of the woods on each of them, and then extend your insects with the threads in a row the same as if singly and loose. Or, thirdly, you can have your extending boards and pieces of wood, so to call them, in one, as it were, combining the excellencies of both: as thus—on the board, made in the ordinary way, namely, a thin piece of cork, which, by the way, you can procure at any good shoemaker's, glued on to a piece of deal, and papered over, add a second piece of cork, fastened on the first in the same manner: round off this top piece in suitable lengths on each side cross-wise, and also cut out a strip in the centre of it, and you then have a series of the rounded woods of cork, on which you can extend the insects you catch, either with thread, or ordinary card brace, and can place them in safety in the case which I proceed to describe.

"FUNDED SECURITIES."

If your extending boards are left lying loosely about, it is ten to one but that some damage will accrue to the specimens that may from time to time be placed thereon. There are various accidents to which they may be exposed, to say nothing of dust, which is an unfailing source of damage and injury. To guard therefore against this, "No quid detrimenti res entomologica capiat," have a case made of oak, or any other wood, say one foot three inches high, one foot one inch wide, and nine inches deep; with a door to it, and inside a series of slides on which the boards can run, so as to be easily taken in and out. I had my extending boards made of narrow width, so as to have two on each tier, sufficient height being left of course between each for the pins of the insects and the card braces, and I have lately had them again further divided into two each, so as to have four of them one inside the other. The advantage of their being narrow is that you can set the insects one after the other in a row, either sideways or lengthways, and you thus avoid the various "moving accidents" which otherwise the setting of one in the way of another exposes each and all to.

"SECOND THOUGHTS ARE BEST."

All that I have said as to the desirableness and necessity of having a cabinet, and that a good one, for the preservation of your specimens, I still keep to; but I have since been made cognizant of another kind of receptacle for them, which is equally good in most respects, though not quite in all, and better in some. The Rev. William Bree, of Polebrook, near Oundle, Northamptonshire, first shewed me this plan. It is to have cases made, such as backgammon or chess boards, resembling large folio books, corked and glazed inside, covered with leather, and lettered on the outside, at least they may be, "as you like it," "British Entomology," "volume i," "volume ii," and so on.

Since I saw the Rev. Mr. Bree's, I perceive that Dr. Baikie, of the Naval Hospital, Haslar, has written about these blank volumes in "The Naturalist," vol. ii, page 207, and, which is better, has told us where they may be procured, well made by a person in the habit of making them, namely, Mr. Robert Downie, of Barnet, Hertfordshire. To him I lost no time in writing for further information, and I give the result to the readers of my "APHORISMATA" —"Aphorismi," by the way, my brother told me it should have been; but as he took his "First Class" at Oxford in 1849, "Term: Pasch:" and I my "Second" so long ago as "Term: Mich:" 1833; when, I may here record, I took up part of "Pliny's Natural History" for the first time in that learned University, to the no small astonishment and discomfiture of the Examiners 'In Literis Humanioribus,' I trust my said readers will pardon me the 'lapsus,' and at the same time this lengthened and somewhat involved sentence.

But 'ad rem:' Mr. Robert Downie furnishes me, and through me my readers, with the following list of apparatus which he is always ready to furnish, and which, as I truly believe they will be found good and useful, as well as cheap, I heartily recommend to all who are desirous of the proper preservation of their specimens I give the whole of his catalogue, as well as the part relating to the books, through the desire to benefit a person who seems to be a deserving man:—

1.—"An improved book box, which excludes the air and dust from the insects, covered with green book cloth, gilt labels, corked top and bottom, sixteen inches by twelve; the same as those made by me for the British Museum, and when shut up they resemble two volumes of a book: twelve shillings each.

2.—The next size, finished in the same style, and corked top and bottom, thirteen inches by nine and a half: seven shillings.

3.—Deal store boxes, corked top and bottom, sixteen inches by fourteen: eight shillings. If made for foreign insects, one shilling extra.

4.—Mahogany collecting boxes, from four shillings and upwards.

5.—A drying safe or box, with four trays corked, a drawer with divisions for pins, perforated zinc front and back, lock and key complete: twelve shillings and sixpence.

6.—An improved whalebone net, which answers all the purposes of sweeping, beating, or for collecting insects on the wing: reduced to twelve shillings and sixpence. It is portable, and shuts up in a case like an umbrella.

Sheets of prepared cork for cabinet drawers, sixteen inches square: two shillings each.

All kinds of boxes and apparatus on improved principles made to order: prices in proportion as stated above."

I need only remark in conclusion that while a cabinet, especially if a large one, is rather an expensive affair, the drawers costing ten shillings each, the books, on the principle of a division of labour, or rather of spreading an expense over a longer time, cause it to be hardly felt. One more last word: I recommend the books to be kept upwards, as if on a shelf and not on their sides, for otherwise that which is a detriment to the preservation of insects in ordinary boxes, will exist here also, namely, the dust will fall from the specimens on the upper side, and lodge on those on the lower one.

"DEATH IN THE BOTTLE."

This is a true motto—one which it behoves others as well as Entomologists to bear in mind; it is, however, only with the latter that I have at present to do. Various opinions have been set forth at greater or lesser length, on the so-difficult-to-be-decided question, what is the amount of feeling that insects possess? Into these I shall not now enter, but shall content myself with enunciating the maxim which I promulged in the "Zoologist," page 1680, namely, "With regard to the feeling of insects, as much has been said, and much may be said, on both sides, I would only beg to add that I think there can be no doubt that, whatever opinion any may form or may have formed on the subject, it will be the best and safest way for all to act on the supposition that they have some, if not a very high degree of feeling, and accordingly to make it an unfailing rule to kill them as instantaneously as possible." To this I still adhere, as will, I hope, all my "gentle" readers likewise; and I have it in my power to make known a simple and efficacious mode of killing lepidopterous—and I doubt not any other insects—if not instantaneously, yet almost so, and certainly, at all events, without any sensation of pain whatever. It is not indeed my own discovery—'nec meus hic sermo est,' but that of a gentleman, Charles Barron, Esq., who has published the first account of the method that I am aware of, in the "Zoologist," page 3435, dating from the Royal Naval Hospital, Haslar, March the 3rd., 1852. His plan however is rather a complicated one, and the following improvement upon it will be found well worthy of your especial attention. The agent to be employed is the well-known—though only recently in this application of it —Chloroform!

Go to a druggist's, and purchase a strong, wide-mouthed, moderate-sized glass bottle, namely, large enough to hold a large moth or butterfly. It should be of one width all the way up, for the reason to be presently mentioned, and should have a glass stopper, so as to make it air-tight, or as nearly so as possible. Fill the bottom of this bottle with sponge, and over the sponge place a piece of perforated zinc, which you will now see you could not do unless the bottle was of a uniform width. The use of the zinc is to keep the wings of the insect from touching the sponge, for it will soon absorb the liquid residuum of the drug, and so might and would wet and injure the wings. When you are going out collecting, or if at home you want to kill any insects that you may have reared or otherwise obtained, pour a few drops of chloroform into the bottle, which will make its way to the sponge through the perforated zinc, and immediately "put the stopper upon it." Take it off when you want to put an insect into the bottle, and then, putting it on again, in a few seconds at the most the insect will be—apparently at all events—lifeless, and that without pain to itself, or injury to it as a cabinet specimen. Observe, however, that, and especially if you have occasion to remove the stopper at all often, the spirit evaporating, will require to be renewed, and you must therefore carry with you a small phial of it, so as to be able to replenish the larger one as often as may be necessary. You will find also that if the bottle be kept upright and not needlessly shaken, a large quantity of insects can be well and safely brought home in it, without being transferred to the pocket-box. This at the same time ensures that care which should always be taken to leave the insects a sufficiently long time under the influence of the narcotic, for otherwise the state of "coma" would go off, and the anticipated result would not be gained. I have seen spirit of ammonia applied in the same way, but the former is decidedly preferable.

"PIN MONEY."

I have already briefly touched on this "head," but it is a subject on which a few pointed remarks may yet be made with advantage, at all events to the "tyro." And, first, where to procure a suitable article is the most important enquiry. If therefore you can find your way about London—which is more than I could do until a few years ago, having never till then even passed through that village—go to Cheapside, a locality of which you have doubtless heard in connexion with Johnny Gilpin, and in divergence therefrom, to the right, as you wend eastward, namely, in a place yclept "Crown Court," you will find the London establishment of a Birmingham firm, by name "Edelsten, Williams, and Co." State your requirements, and for a due consideration, they will be instantly supplied. You must, however, if you write from a distance, accompany your order with a post-office one, and you must also specify that what you want are for entomological purposes, and indeed write the word "Entomological" on the outside of your letter, or otherwise it will be forwarded to the head-quarters at Birmingham, which will cause a delay; but if thus marked, it will be opened in London. I recommend the following sizes as the best, as affording an ample selection for all ordinary uses, but there are many others, the manufactory being of lace pins, and, as such, not expressly for entomological appliances, though they are as good for these as any that I know. The numbers are as follows:—No. 11, price sixpence an ounce; No. 13, sixpence; No. 5, one shilling and threepence; No. 17, two shillings and ninepence; No. 15, three shillings and threepence; No. 18, three shillings. Of these, the largest-sized are only for the largest-sized Sphinges:— for by all means you should incline to putting a too small in preference to too large a pin into any insect; or for putting out the antennæ with; being long and fine, and larger ones being incommodious for the small interstices that will sometimes be found to be left among the card braces.

In no case use any but entomological pins. "A maxim worth remembering I assure ye."

"NOTHING LIKE GLASS."

It will perhaps have been observed that in the description of the entomological book-boxes just spoken of, no mention was made of glass frames to them, which however it is absolutely essential, a 'sine quâ non,' that they should have; or otherwise, every time they are opened, they are liable to all the ordinary injuries from dust and other sources of evil, to which any common entomological boxes are exposed. I wrote therefore to Mr. Downie for a further estimate, which he has supplied as follows, as a postscript to the former particulars:—

A five-shilling box with the addition of glass and frame, would be seven shillings and sixpence; a seven-shilling box with the like would be ten shillings and sixpence, or in other words, half-a-guinea; a twelve-shilling box with the like, sixteen shillings.

The glass to be the very best that can be had, flattened as for picture frames.

"TO BE CONTINUED."

"The most valuable discovery of modern times"—to the Entomologist—is the "Applicability" of sugar to the capture of moths. The "Suggestive Hint" to this mode of proceeding was doubtless furnished to some thoughtful mind, by the observation of the fact, that insects of various kinds resort to an empty cask in which the "Essence of slave" has been placed, for the purpose of "sipping the sweets." Certes, the success that, at least in some places, and on some occasions, "at the season of the year," attends on this experiment, is, as Dominie Sampson used to remark, "Prodigious! Prodigious!! Prodigious!!!" Go the wood, and there, if not shot by some truculent gamekeeper, in whose eyes you will certainly cut a strange and very suspicious figure, and who will have no notion that you are prowling after "untaxed and undisputed game;" there, I say, with a reflector lamp at your girdle, and a flat brush in your hand, wash on a small portion of the trunks of an 'ad libitum' number of trees, the "lotion" I shall presently describe the component parts of. Take of the coarsest brown sugar you can purchase, one pound; of beer, say one pint; boil both well together, and add a little of the liquor which, "My dear young friend, if there be one liquour less abominable than another, it is that commonly called—rum," each time when used. This decoction will be found wonderfully attractive to moths; and on returning to the trees after a proper interval, during which the darkness has come on, you will frequently have both quantity and quantity to choose from. You can best carry the seductive draught in a tolerably large "Pocket Pistol," made by some handy tinman for the purpose—a sort of large "quaigh," with an extra case for its lower part, taking on and off the outside—into which a necessary portion of the mixed ingredients can be poured as required.

"RELAXATION."

I HAVE before alluded by anticipation to this part of my subject, and now proceed 'in medias res'—relaxation you will find it for yourself, as well as for your insects, in the winter time, when you have perhaps—though it is a thing I never have at present—a little leisure on your hands, in which to set your summer captures to your eyes' content. Thus too, having them flexible before you, you, as it were, "fight your battles o'er again," and can indulge the "flights" of your entomological "Fancy" to any extent.

Purchase, which you may do for somewhere about a shilling, at a druggist's, a large glass jar, say one foot high, and six inches in diameter. Let it, if possible, have a glass stopper, which, if you live anywhere near a glass manufactory, you can easily have made, or, if not, a well-fitted cork one, which you can procure at a cork-cutter's. The mouth of the jar is to be as nearly as possible of the same width as the rest of it. Fill this glass jar with plaster of Paris up to a sufficient height, namely, so far as to leave sufficient depth for any insects you may want to relax between the surface and the cork. Pour on the plaster as much water as it will absorb, any surplus being poured off, and over it place a round piece of perforated zinc, the size and shape of the orifice of the jar, the object being to keep the wings of the insects from touching the wet surface; and at the same time the holes in the zinc answer admirably for putting the pins of the insects into, so that they are steadied, and in a manner fixed, so as not to shake about if the jar is moved.

When you have thus placed the said insects in this way, put the cork or stopper tightly into the jar, and in a few hours, more or less according to the size of the specimen, you will have them excellently relaxed, and that without the slightest detriment or damage; the down being as perfect as if they had never been subjected to any such process at all. A tin canister will answer the purpose, but not so well; the jar, especially if it have a glass stopper, being so much more air-tight, and the moisture being therefore the more confined. If a cork be used, it should have a piece of fine kid leather round it, to make it fit close; also, I recommend a basket-work case to guard the lower part of the jar.

"BOTANICAL SPECIMENS."

THE "saccharine juices" of the following plants and trees when in bloom are more or less attractive to moths, and may be therefore cultivated for the purpose, as well as for their respective merits, or examined in their wild state:—

Woodbine, or Honeysuckle, (*Lonicera Periclymenum.*)
Valerian, (*Valeriana rubra.*)
Petunia, (*Petunia violacea* and *nyctaginiflora.*)
Phlox, (*Phlox paniculata, suaveolens, etc.*)

Aaron's Rod, (*Solidago virgaurea*.)
Hop, (*Humulus Lupulus*.)
Nettle, (*Urtica dioica*.)
Pink, (*Dianthus caryophyllus* and *Chinensis*.)
Ivy, (*Hedera Helix*.)
Traveller's Joy, (*Clematis vitalba*.)
Wild Thyme, (*Thymus serpyllum*.)
Barberry, (*Berberis vulgaris*.)
Raspberry, (*Rubus idæus*.)
Pansy, (*Viola tricolor*.)
Common Sage, (*Salvia officinalis*.)
Candied Tuft, (*Iberis umbellata*, etc.;
Sweet William, (*Dianthus barbatus*.)
Lime, (*Tilia Europæa*.)
Jessamine, (*Jasminum officinale*.)
White Verbena, (*Verbena ———? var: fl: albo*.,
Sweet Scabious, (*Scabiosa atro-purpurea*)
Thistle, (*Carduus, Cnicus, var: sp.*)
Laurel, (*Prunus Lauro-cerasus*.)
Privet, (*Ligustrum vulgare*.)
Misseltoe, (*Viscum album*.)
French Marigold, (*Tagetes patula*.)
African Marigold, (*Tagetes erecta*.)
Michaelmas Daisy, (*Aster Tradescantia*.
Blackberry, (*Rubus fruticosus*, *etc*.)
Flote Meadow Grass, (*Glyceria fluitans*.)

"THE DIGGINGS."

IN the winter time, when the insects of the Lepidopterous orders have all but entirely disappeared, their successors of the following generation are to be found in the chrysalis state, or as I recently heard it called "the crystallized state," in various situations; but chiefly at the roots of trees, and especially in those retired "nooks and corners" which afford the most shelter from the severities of the brumal season, and where accordingly the soil is loose and crumbling, and easily entered by the descending caterpillars. By procuring these you frequently obtain several of the rarest species, which otherwise the sight of would never gladden your entomological vision.

You should have an implement made for the purpose, of iron, not too sharp, but sufficiently so, and of any spade-like shape that may most recommend itself to your judgment. Many insects are at the same time to be found in the moss on the trunks of the trees; and these you can look for at more leisure, if you bring a quantity of the moss home with you, and examine it carefully over a sheet of white paper; giving liberty to any specimens you do not want, and throwing away the moss again, with the like intent towards any that it may yet contain.

"THE PROCREANT CRADLE."

I RECOMMEND a large square or oblong case, a foot, or a foot and a half in diameter, more or less as the case may be, and proportionately deep, for keeping the chrysalides in, which, during the summer, autumn, or winter, you may have collected together. The case should be made of fine wire netting, so that you can see into it, and discover anything that may have come out; and at the same time of close texture, so that it may not come out, in a different sense of the words.

The object of its being thus large is, that during any interval of time which may elapse, the insects may not damage themselves, as they might do in a more confined space. I also recommend a little moist sugar being kept in the case, on which, if they choose, they may feed.

"OMNIUM GATHERUM."

WHEN you have several thoughts in your head at one and the same moment, it is somewhat difficult to retain them all sufficiently long to commit them to paper—a few "random recollections" therefore I now proceed to indite.

The pieces of wood for the extending boards should be of an uniform shape, in having the curved part tapering for the same length—the intermediate part between it and the centre being flat, or nearly so. They may then be all of exactly the same height, and the corners all squared off to exactly the same depth.

The following sizes are those I have, on mature deliberation, determined on as the best for myself, and therefore for all other entomologists:—Five inches and a half, five inches, four and a half and four, three and a half and three, two and a half and two, one and a half and one; and, observe, these measurements are from the top of the side cut off to its opposite; that is to say, the clear part on which the wings can be extended.

The groove in each of these pieces of wood should be proportionate to the size of the bodies of the insects for setting, which each size of wood is intended; namely, the smallest-sized piece of wood should have the smallest groove, the largest-sized the largest, and so on with the intermediate ones in gradation. Let so much suffice for this mode of extending insects.

In relaxing insects to remove a bad or too large pin, or to remove such when relaxed for any other additional purpose, do not push the insects downwards towards the point of the pin, (which is to be done against any hard surface,) but press it upwards, at least first, towards the head of the pin, and then when once shifted it is easily taken out altogether. Otherwise in pressing downwards against the thorax of the insect, you can hardly fail to injure the down.

The pieces of silver paper are first to be made of a square shape, and then one corner should be torn off, which part should be placed against the base of the fore wings, and thus they will be found to lie better upon them, and be more readily kept in place by the first windings of the thread. "Crede experto."

Another advantage of the silver paper is, that if the wings be completely covered with it, as they should be, no dust can accumulate upon them previous to the insects being placed for safety in the cabinet. This is no small advantage, for, even if ever so carefully kept otherwise, some amount of the evil so much to be guarded against cannot but befal.

If you are at all in haste to have any relaxed specimens dried, you can accomplish the object by placing them within your fender, namely, if the fire in the grate be lighted; but observe, for good effect in this process, the wings should be completely covered and well held down in every part with the silver paper, as otherwise they might and would spring up out of place here and there, in a manner the very reverse of desirable.

Item.—They should not be taken off too soon, but to speak scientifically, the caloric absorbed should be suffered gradually to evaporate; in plain English, they should be left untouched till quite cool again.

But—Memento—beware of sparks, and do not stir your fire while the insects are drying before it. If you are a lady, (and I am happy to know that there are entomological ladies, and happy to think that there may be some such among my readers,) take my advice, and never stir the fire at all. There are several good and weighty reasons for this advice; one is, if I may be pardoned for saying so, that no woman ever could stir a fire. Do not, therefore, try—be content with excelling men in many important particulars; and "assure yourselves of my high consideration."

N.B.—Do not dry recent specimens in this way.—Cardinal Mazarin's motto, "I and Time," should be yours, as it is mine—practically speaking.

"TRANSPARENCIES."

I have found by experience that the silver paper over the wings presents one hindrance, namely, that it prevents your seeing through it to ascertain whether on their final adjustment the wings are perfectly even on both sides or not. To remedy this defect, I have adopted the expedient of procuring transparent silver paper, through which you can see sufficiently well for the purpose; it should of course be the thinnest and finest that can be obtained. It is to be had of any bookseller, and is called Tracing paper; you can, if necessary, make it for yourself, by slightly oiling common silver paper.

"CLAPTRAP."

As I have before said, the large net which I have already described is by far the best for all ordinary purposes; others, however, may be used with more or less advantage; and a "Sweeping machine" is necessary for obtaining Water-beetles, and those insects, both Coleopterous and other, which are procurable among the long grass on the bank or hedge-side. It may be made as follows:—

Get a good strong walking-stick, which will be often found useful in more ways than one, and have a small round strong wire rim or hoop made to fit to the end of it with a screw. To the rim attach a pretty strong canvass net of any convenient size, say two feet in diameter, and a foot and a half in depth. You can procure the whole complete "for a consideration"—say some five shillings. By reversing it every now and then, you get rid of all odd-ments that you do not want, for "all is not fish that comes to the net."

To catch the Purple Emperor, and other flies that frequent the tops of the highest trees, you want a small round net with a handle some fifty or sixty feet long. This is a very difficult thing to accomplish, and it is but an un-wieldy implement when made in the ordinary way, but I have built a castle in the air in the shape of a very long fishing-rod, made of light bamboo, all the upper part to be kept from bending and breaking by means of shrouds or stays coming from near the top to a double cross-tree, like those from the royal mast of a man-of-war. An illustration will be given with the other engravings, and I hope to find it as effective in practice, as it is in theory on paper. The net is to be made of very light open net-work, so as not to catch the air.

"WHAT'S IN A NAME?"

A good deal too much in many an entomological one. Hardly two "Lists" agree; and, as I can recommend no one in existence, I must briefly dismiss the subject of nomenclature by expressing the hope that eftsoons we may be permitted to resort to the primitive simplicity whilom enjoyed by our ento-mological forefathers, and that an insect yclept by one name may be deemed to be sufficiently denominated, so that the pride of nomenclators may be no longer fostered by dubbing their unconscious adoptions with as many titles and "family" distinctions as would suffice a Spanish grandee, to say nothing of their unpronounceable barbarisms, which offend against all laws of classical propriety. I have myself used Mr. Doubleday's catalogue, price half-a-crown, published by Mr. Voorst, London. It is well printed on good strong paper, and only on one side, so as to be able to be cut out for labels for cabinets, and also for marking in the species possessed.

"DE OMNIBUS REBUS ET QUIBUSDAM ALIIS."

I HAVE almost exhausted the previous part of this wide subject under the former heads, but I must endeavour to say something under the latter part of it; and, first, I may add, that in order to keep the wings sufficiently down with the silver paper, two or even three pieces will sometimes be required on each side. Further, the mode just mentioned of drying the wings before the fire, will be found very effective in keeping them permanently in the way they are placed, so that they may be effectually as well as nominally "set."

A very good method of procuring many rare Lepidopterous, and indeed other insects, is by shaking in the day-time any young trees which may admit of such an effect, the result being to dislodge those which may be resting under the leaves, from whence they either fly or drop down into the grass beneath or at some little distance. In this way, some years since, I procured a very large numbers of splendid specimens of the *Tryphæna fimbria*, then thought a very rare and valuable insect, and in "The Naturalist," old series, volume ii, pages 83-4-5, I gave an account of the whole mode of procedure and its results, recording how, in plain prose, when divers entomologists adopted the plan, which, as far as I know, was the invention of Mr. Hugh Reid, of Doncaster, the coppices resounded again with the "kicks of the sturdy entomologists"—poetically speaking—"how bowed the woods beneath their sturdy stroke."

Another means of relaxing specimens, though, in my opinion, by no means so effective as that hereinbefore described, is by means of bruised laurel leaves. The following is the method adopted, as given in the "Zoologist," pages 1343-44:—

Mr. J. W. Douglas, of 6, Grenville Terrace, Coburg Road, Kent Road, London, writes, "A quantity of laurel leaves, (thirty or forty,) is much bruised, put into a bag, and enclosed in an air-tight vessel; on the bag are placed the insects wished to be relaxed, and they become flexible in a few hours more or less, according to their size. The advantages of this system are, that the insects may be left for any length of time without getting mouldy, and that moths of a green colour or delicate texture may be operated upon without injury, none of which were possible on the old plan. It is somewhat singular that this relaxing effect should be produced by laurel leaves, which contain a large amount of prussic acid, because if an insect be killed by that poison, its membranes become intensely rigid."

In the following article, Mr. Samuel Stevens, of 38, King Street, Convent Garden, London, says in like manner, "Through the kindness of Mr. Dale, I have been informed of a most excellent method of relaxing Lepidoptera and other insects, and having adopted it lately, and finding it answer uncommonly well, I think it will be a great benefit to entomologists to make the plan generally known. I procure about a dozen shoots with the leaves of the common laurel, the younger the better, put them into a coarse bag or cloth, (a shot-bag I use,) bruise them well with a wooden mallet till the bag becomes quite moist, then put it into a jar or other wide-mouthed glass vessel, and stick the insects on the top of the bag, which must be tied over or secured in some way, so that it be made perfectly air-tight. Twenty-four hours are generally sufficient to relax most insects, but one great advantage is, that if they remain a week or ten days in the laurel they are not the least injured, so that they can be set out at any convenient opportunity; it also completely destroys the mites or mould if the specimens be infected, and it will be found to have a great many advantages over the old plan of damp sand. I was in hope, from experiments that I made on two or three green species, that the colours would not fly; but since I regret to find, on further trial, that *Hipparchus papilionarius*, *Hemithea vernaria*, and *Hemithea cythisaria* are considerably changed by it. Mr. Dale informs me that it answers equally well with the other orders:—he having relaxed nearly the whole of his Dragon-flies, and it is much used at Bristol for the *Hymenoptera*."

Postscript.—The following items are extracted from Mr. Edward Newman's "Familiar Introduction to the History of Insects:"—" The Entomologist should be provided with two wide-mouthed vials; one empty and perfectly dry, having a quill passing through the cork, and going a considerable way below it: this quill may be stopped at top by a second small cork: within the vial some blotting-paper may be kept, which not only absorbs any moisture, but serves as something to crawl on for the living insects which are taken from time to time and dropped through the quill. The other vial should be made very strongly, well corked, and three parts filled with spirit; common whisky is the best; pure alcohol injures the colours." "Quills cut off close to the feather are very useful for bringing home minute insects of all classes. The aperture should be most carefully corked, the corks being cut expressly for the purpose, and should be of sufficient length to go half an inch into the quill, and thus not liable to come out in the pocket."

The following, by Mr. T. B. Hall, of Woodside, Liverpool, is from "The Naturalist," old series, volume iii., page 159.—" Substitute for Cork Lining in Entomological Cabinets.—Having forwarded the receipt committed to you by Mr. Morris to a very excellent Entomologist of Liverpool, A. Melly, Esq., for the purpose of asking his opinion respecting it, he states that he has always been in the habit of using composition instead of cork, and that he finds it not only cheaper, but quite equal to cork, and that on the Continent the plan is generally adopted. The one he employs is much harder, and is composed of two thirds of the best bee's-wax and one third of the best resin; but he observes that, in this climate. the addition of tallow cannot do much harm, and will save something in the cork: the great point is to melt it well, and to pass the resin through a sieve before the wax is added."

The pins you want to take out with you when collecting, to put through any insects you have "netted," after they have been perfectly killed, may be carried either in the pocket-box or in a small thin pincushion, attached to a "guard." Two of these, made of velvet, and exactly resembling butterflies, have been presented to me by Dr. Henry Moss, of Appleby: I give a figure of one.

One more last word: it has occurred to me that by driving a tin tack firmly, but not up to the head, on each side of the rounded pieces of wood, they may, after the insects have been set upon them, be firmly lashed on to the narrow extending boards by a twine wound underneath them, as illustrated in the engraving, and thus they may be carried safe in the setting case without being liable to be shaken about. Any respectable draper can procure the proper thread for setting the insects with, from the Messieurs J. and W. Taylor, Leicester, and "made to order," wound singly for the purpose.

And now I have given you, and I think sufficiently, natheless not at an undue length, the results of an experience of many years standing. I was born an Entomologist, was self-educated one, as the cabinet in the Ashmolean Museum at Oxford, which I found time, amongst other multiform and deep studies, while there, to arrange, will testify; and it is nothing but the more serious business of life that now in great degree hinders a larger amount of the innocent enjoyment which the science of Entomology so abundantly rewards her votaries with. "Valeas," good reader, and may you never "go out" without catching a Purple Emperor, or a Scarce Swallow-tail, a Large Blue, or a Pale Clouded Yellow, a White Admiral, or a Camberwell Beauty, and if these pages shall have assisted you in the chase—"Plaudite."

ADDENDUM.—To the list of plants attractive to Moths, add the Sweet Willow, the Larkspur, the Bladder Campion, (*Silene inflata,*) the Reed, (*Arundo,*) and the Sallow, (*Salix.*)

NOTA BENE.—Some kinds of wood are very injurious to specimens contained in cabinets made of them; oak and mahogany are the best. I have known a good collection much injured by being kept in a cabinet made of ash or elm, I forget now which: turpentine exudes into the drawers, and is very prejudicial.

Lastly, in common with all who wish well to their collection or to their country, I deprecate frequent "Changes in the Cabinet"—"Let well alone" is a good and wholesome proverb, applicable both politically and entomologically.

FINIS.

www.ingramcontent.com/pod-product-compliance
Lightning Source LLC
Chambersburg PA
CBHW031857220426
43663CB00006B/660